AF510867

Springer
Berlin
Heidelberg
New York
Barcelona
Budapest
Hongkong
London
Mailand
Paris
Santa Clara
Singapur
Tokio

Ciarán Ó Fágáin

Stabilizing Protein Function

Springer

Ciarán Ó Fágáin, Ph. D.

Dublin City University
School of Biological Sciences
Dublin 9
Republic of Ireland

ISBN 3-540-63189-5 Springer-Verlag Berlin Heidelberg New York
Biotechnology Intelligence Unit

Library of Congress Cataloging-in-Publication Data

Ó Fágáin, Ciarán.
Protein stability and stabilization of protein function / Ciarán Ó Fágáin
p.cm. - (Biotechnology intelligence unit)
Includes bibliographical references and index.
ISBN 1-57059-461-9 and 3-540-63189-5 (alk. paper)
1. Protein engineering. 2. Proteins–Stability. 3. Proteins–Denaturation. I. Title. II. Series.
[DNLM: 1. Protein Engineering. 2. Proteins–chemistry. TP248.65.P76F151p 1997]
TP248.65.P76F34 1997
660'.63–dc21 97-17143 CIP
DNLM/DLC
for Library of Congress

This work is subject to copyright. All rights are reserved, whether the whole or part of the material is concerned, specifically the rights of translation, reprinting, reuse of illustrations, recitation, broadcasting, reproduction on microfilm or in any other way, and storage in data banks. Duplication of this publication or parts thereof is permitted only under the provisions of the German Copyright Law of September 9, 1965, in its current version, and permission for use must always be obtained from Springer-Verlag. Violations are liable for prosecution under the German Copyright Law.

© Springer-Verlag Berlin Heidelberg and Landes Bioscience Georgetown, TX, U.S.A. 1997
Printed in Germany

The use of general descriptive names, registered names, trademarks, etc. in this publication does not imply, even in the absence of a specific statement, that such names are exempt from the relevant protective laws and regulations and therefore free for general use.

Product liability: The publisher cannot guarantee the accuracy of any information about dosage and application thereof contained in this book. In every individual case the user must check such information by consulting the relevant literature.

Typesetting: Landes Bioscience Georgetown, TX, U.S.A.

SPIN:10631277 31/3111 - 5 4 3 2 1 0 - Printed on acid-free paper

Dedication

Do mo dhá chlann: an ceann inar tógadh mé agus an ceann ar chabhraigh Margaret liom a bhunú.

'When I am dead
I hope it may be said
His sins were scarlet
But his books were read.'
- Hilaire Belloc

PREFACE

Protein stability is an issue of both scientific and practical interest. A protein for study or investigation must remain (or be kept) stable during isolation and purification procedures. Since many commercially-important biomolecules are proteins, the stability and stabilization of proteins is of crucial importance in biotechnology and especially in medical biotechnology. From a development or production viewpoint, stability dictates the operating lifetime and/or shelf life of protein-based bioreactors, therapeutics, diagnostics or biosensors. From a regulatory viewpoint, protein stability can have a critical bearing on product efficacy, shelf life or declared activity and any stabilization strategy must accord with regulatory provisions currently in force. Limiting stabilities of crucial proteins continue to act as a brake on the development of the biotechnology industry.

There is a need for a concise book describing the fundamentals of protein stability and its manipulation/improvement by a number of means. The present volume aims to satisfy this requirement. It deals principally with in vitro stability and stabilization of proteins, whether they act as bioprocess components, therapeutics, diagnostics or in biosensors or other analytical devices. It pays little attention to the in vivo stability of endogenous (or exogenous) proteins or on the turnover of cellular proteins.

This book emphasizes protein stability as the persistence or retention of biological activity, not just as folding stability. Its focus and presentation reflects experiences in biotechnology industries and in academic life. It looks at a protein as a working tool (catalyst, sensor, therapeutic) rather than as just a molecule for study. Emphasis is placed on the deliberate treatment of target proteins to increase their tolerance of adverse conditions, and often the focus is on practical examples with biotechnological or medical applications. Stability parameters are outlined and current understanding of stability (arising from studies on extremophilic proteins, proteins in organic solvents, deleterious reactions and stabilizing interactions) is reviewed. The book goes on to describe strategies for the manipulation of stability, some of which are empirical and deceptively simple, while others involve the use of leading-edge knowledge and techniques. It describes the manipulation of stability by the use of additives, chemical modification of proteins in solution, immobilization and protein engineering. It compares the merits of these alternatives and attempts to draw general principles from empirical successes. It also discusses long-term storage and includes numerous case studies. There are copious references to the original research papers and to relevant reviews/monographs on particular topics. Much of the literature on stability and stabilization has been concerned with relatively simple and well characterized proteins, for fairly obvious reasons. While I have included much material of this sort, I have tried wherever possible to pick industrially or medically relevant examples and case histories. Also, much attention has been paid to those stabilization strategies and techniques which seem to have the greatest potential for therapeutic or other "medically-applicable" proteins. In emphasizing these points, this volume differs from many of those currently available.

Part 1 opens with a chapter outlining models of protein stability and reviewing the main protein stability parameters. Chapter 2 shows how an understanding of protein stability has emerged from investigations of protein denaturation and inactivation and from detailed studies on stabilizing interactions in proteins. Subsequent chapters review the stability characteristics of enzymes in nonaqueous media and the special adaptations of thermophilic proteins. Part 2 describes the manipulation of protein stability in successive chapters on the use of additives, chemical modification of proteins in solution, immobilization and protein engineering. A separate chapter considers long-term storage of proteins.

It is anticipated that this book will interest a variety of academic and industrial scientists, notably protein scientists, enzymologists, biotechnologists, those involved in quality control/stability testing of natural or recombinant proteins, diagnostic kits, therapeutics, enzymes (especially enzymes applied in the detergent, food, fermentation, fabrics and other industries). It will also be of interest to students, especially those at postgraduate level. Students of biotechnology or of life sciences will find the text useful as a general survey of the field, for information/updates on specific topics, or as a core or reference text in the laboratory. The book does not consider the special requirements of pharmaceutical regulatory authorities, good manufacturing practice (GMP) or ISO 9000 specifications. Readers who need to meet these directives and standards should consult materials produced by the appropriate authorities in order to ensure compliance.

Acknowledgments

I wish to thank sincerely a number of scientists who assisted me (as did some companies) in various ways. Their assistance included helpful correspondence, the supply of preprints of articles (or copies of just-published material) and permission to reproduce copies of previously-published material. This list includes Amicon Inc. (P. Schratter and T. Elwin), F.H. Arnold, K. Breddam, P. Bryan, S. Doonan, F. Franks, Gelman Scientific Inc. (B. Dawson), J.M. Guisán, A. Haselbeck, Heto A/S (S.-E. Meelby), R. Jaenicke, A.M. Klibanov, P.L. Mateo, B.W. Matthews, O. Misset, V.V. Mozhaev, C.N. Pace, R. Rudolph, J.A. Schellman, L. Serrano, K.G. Welinder, J.A. Wells.

Particular thanks are due to some friends and colleagues at Dublin City University: to Richard O'Kennedy for contributing to the antibody section of chapter 8 (pages 141-144), Ellen Breen, Science and Engineering Librarian and Tony Killard who contributed Figure 8.6. I wish also to thank my graduate students past and present: Fiona Manning, Enda Miland, Ann Murphy, Damien O'Brien, Anne-Marie O'Brien, Orlaith Ryan. I must acknowledge also the patience and tolerance of my wife Margaret and children Olga and David during the many evenings I have spent at the PC console. Finally, I should thank God, since my completing this book is little short of miraculous.

Ciarán Ó Fágáin
March 1997

CONTENTS

Part 1:
Understanding Protein Stability

CHAPTER 1

Protein Stability and Its Measurement

Introduction

Enzymes are protein molecules that act as extremely efficient catalysts. They can achieve rate enhancements in excess of 10^{12} compared with the uncatalyzed reaction and demonstrate notable molecular and even chiral specificity. Despite these favorable properties, the usefulness of enzymes and proteins as analytical tools and as industrial catalysts is often limited by their requirements for 'mild' storage and reaction conditions. This is because many enzymes lose their catalytic abilities over time, that is, they have poor operational or long-term stability. Stability is an issue also in the development and use of protein-based analytical or sensor devices, and it can be of literally vital importance in protein pharmaceuticals or therapeutics, where deterioration of protein preparations over extended storage periods (i.e. a limiting 'shelf life') can be a serious drawback.

Protein stability is, therefore, a very important area of study within biotechnology. Interest in this topic will likely grow due to the increasing use of recombinant therapeutic proteins, the advent of protein engineering and recent strides in understanding protein folding. In order to discuss this important topic of protein stability, one must be able to define it and to measure it quantitatively. This is THE purpose of chapter 1. A different but related issue is the loss of activity occurring during isolation of the target protein. Ways of minimizing this latter problem have been outlined by Volkin and Klibanov.[1]

Protein Structure

A protein, or polypeptide, is a linear, head-to-tail polymer of L-alpha-amino acids (see Fig. 1.1). Twenty amino acids with different R-group side chains are commonly found in proteins. The carboxyl group of amino acid 1 is linked via a *peptide bond* to the alpha-amino group of amino acid 2 and so on. Each protein has therefore an *amino-* or *N-terminus* and a *carboxyl-* or *C-terminus*. The sequence of amino acids in any protein is dictated by the linear information on the DNA of the gene coding for the protein. The protein is assembled step by step, one amino acid at a time, on the ribosome with the involvement of messenger and transfer RNAs.[2]

This linear sequence, however, gives rise to more complex orders of structure. The groups involved in peptide bonds form hydrogen bonds with other appropriately-spaced peptide bond participants. Regular *secondary structures* arise from

Stabilizing Protein Function, by Ciarán Ó Fágáin.
© Springer - Verlag and Landes Bioscience 1997.

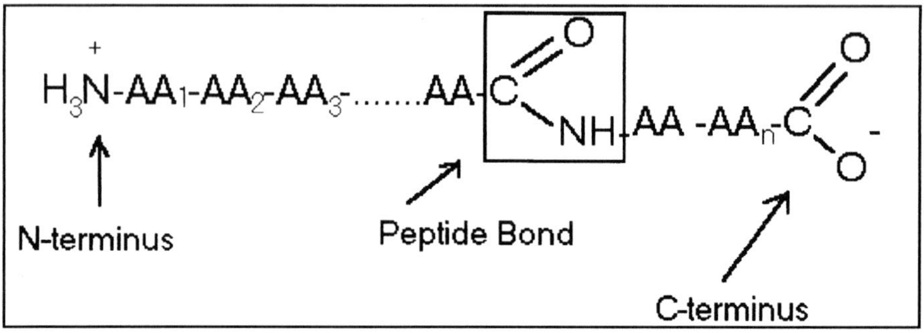

Fig. 1.1. Linkage of amino acids to form a polypeptide.

these bond networks. Chief among these structures are the *alpha-helix* and *beta-sheet.* Meanwhile, the different side-chain R-groups attached to the alpha-carbon of each amino acid form various types of bonds (hydrogen, ionic and disulfide bonds, hydrophobic and van der Waals interactions), giving rise to a three-dimensional protein *tertiary structure.* Occasionally the fully-formed protein will consist of two or more polypeptide chains associated with one another. The number and nature of these chains (dimer, trimer etc. of similar or dissimilar subunits) is referred to as the *quaternary structure* of the protein in question (Fig. 1.2). Proteins often undergo *post-translational modifications* such as proteolytic processing (e.g. removal of an N-terminal signal sequence), glycosylation or the modification of a glutamic acid residue to give pyroglutamic acid. More details on these processes, together with a much fuller description of protein structure, may be found in ref. 3.

It is an axiom in biology that structure relates to function. Integrity of the three-dimensional tertiary (or quaternary) structure is essential for the correct functioning of the protein, e.g., catalysis of a reaction by an enzyme, or antigen binding by an antibody. Loss of quaternary or tertiary structure leads to loss of function. Adverse conditions of temperature, pH or solvent, or high concentrations of urea or guanidinium hydrochloride, can bring about this loss of function. Heavy metals, certain organic chemicals or chelating agents can act similarly in some cases.

Definition and Measurement of Protein Stability

The term *stability* refers to a protein's resistance to adverse influences such as heat or denaturants, that is, to the persistence of its molecular integrity or biological function in the face of high temperatures or other deleterious influences. A perfectly folded, fully functional monomeric protein can lose its biological activity in vitro by *unfolding* of its tertiary structure to a disordered polypeptide, in which key residues are no longer aligned closely enough for continued participation in functional or structure-stabilizing interactions. Such unfolding is termed *denaturation.* It is usually cooperative and may be reversible if the denaturing influence is removed, since the polypeptide chain has not undergone any chemical changes[4] (see Fig. 1.3).

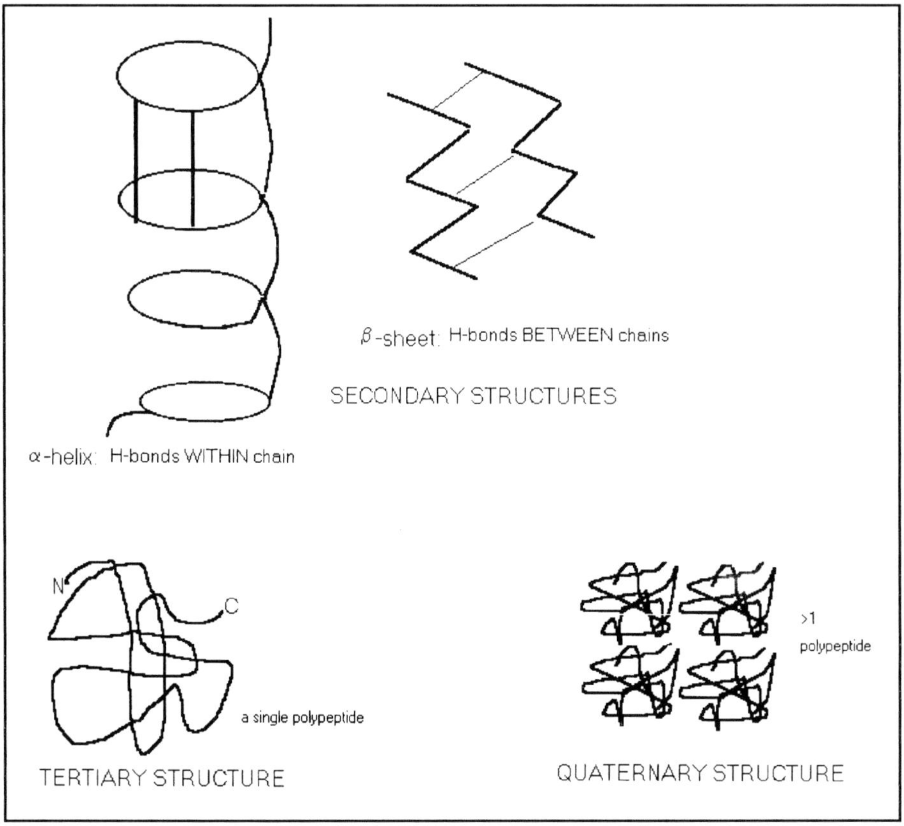

Fig. 1.2. Protein secondary, tertiary and quaternary structures.

A protein is also subject to chemical changes which lead to an irreversible loss of activity or *inactivation,*[4] particularly following unfolding (Fig. 1.3). An unfolded, extended polypeptide will be much more prone to proteolysis than a tightly-packed, globular protein.[5] Unfolding may result in the loss of a functionally-essential co-factor from a holoprotein, such that biological activity will not be regained even if the unfolding can be reversed to yield the corresponding apoprotein. Unfolded polypeptide chains may aggregate to form an inactive, insoluble mass while an individual chain 'attempting' to refold may enter an incorrect, kinetically-trapped conformation from which it cannot emerge. (A living cell may be able to prevent these events. Cells contain a number of so-called chaperone proteins which assist the folding of newly-synthesized proteins in vivo.[6] Chaperones can also act in vitro to prevent aggregation and assist folding.[7,8] At least one such protein has been shown to reactivate an aggregated enzyme.[9]) Peptide bonds and amino acid side chains are chemically reactive and can participate in deleterious reactions, particularly at high temperatures or under inappropriate storage conditions. Many of these latter processes are now quite well understood (see 'Deleterious Reactions' in chapter 2).

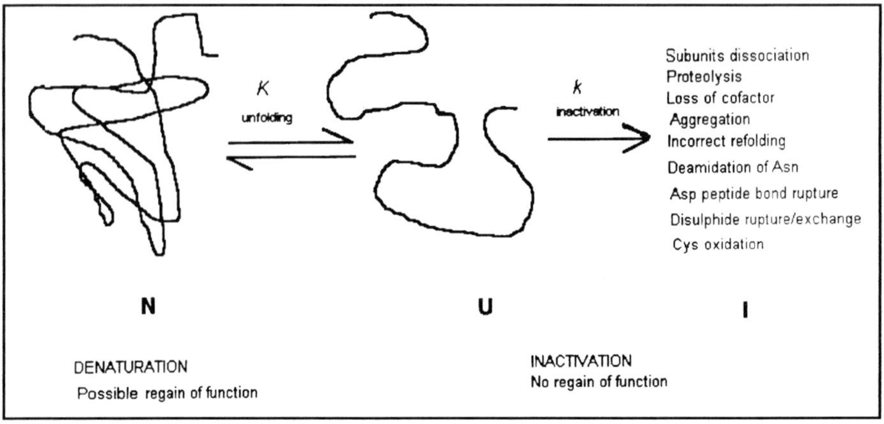

Fig. 1.3. Denaturation and inactivation of a protein molecule. N, native protein; U, unfolded protein; I, inactivated protein. Reprinted from: Biochim Biophys Acta 1252, Ó Fágáin C. Understanding and increasing protein stability, 1-14, 1995, with kind permission of Elsevier Science - NL. © 1995 Elsevier Science - NL, Sara Burgerhartstraat 25, 1055 KV Amsterdam, The Netherlands.

These different molecular phenomena give rise to two distinct definitions of in vitro protein stability. These are *thermodynamic* (or *conformational*) stability and *long-term* (or *kinetic*) stability. Thermodynamic stability concerns the resistance of the folded protein conformation to denaturation (i.e. its Gibbs, or free, energy of unfolding) while long-term stability measures the resistance to irreversible inactivation (i.e. persistence of biological activity). Both types can be represented in a single scheme:[10,11]

$$N \xleftrightarrow{K} U \xrightarrow{k} I$$

where N stands for the native, U for the unfolded (denatured) and I for the irreversibly inactivated forms of the protein. K is the equilibrium constant for the reversible $N \leftrightarrow U$ transition while k is the rate constant for the irreversible $U \rightarrow I$ reaction. The $N \leftrightarrow U$ transition corresponds to thermodynamic stability while the $U \rightarrow I$ reaction represents long-term stability.

Folding Stability

Folding (conformational) stability is usually measured by optical techniques (such as UV spectrophotometry, fluorescence or circular dichroism) or by urea gradient gel electrophoresis. Many scientists will be familiar with these techniques and there are useful practical descriptions of their use in refs. 12 and 13, respectively. Recent reports describe the analysis of thermal denaturation by free solution capillary electrophoresis[14] and temperature gradient gel electrophoresis.[15] These methods are sensitive to changes in protein conformation and thus monitor unfolding of the target protein. Free solution capillary electrophoresis achieves molecular separations without mass transfer between mobile and stationary phases and without retardation by a solid gel matrix. The separations occurring, therefore, depend on the intrinsic properties of the sample but accurate measurement of temperature is essential. The technique allows estimation of the apparent thermodynamic parameters (Gibbs energy, enthalpy and entropy; see below) as well as T_m (see below) and, uniquely, can show the population distribution of mobility states. It is reportedly as accurate as differential scanning calorimetry in the determination of apparent thermodynamic parameters.[14] Temperature gradient gel electrophoresis involves the use of two electrode tanks and a horizontal plate with two water baths connected to provide a temperature gradient perpendicular to the direction of the electric field. The gel is connected to the electrode tanks via wicks. One of the water baths should have a refrigerator to maintain a suitable low temperature. In this system, bovine serum albumin showed discontinuous bands on melting, indicating an irrreversible thermal transition. The apparent T_m (see below) was 58°C in 2 M urea. These findings were consistent with a thermal denaturation of bovine serum albumin monitored spectroscopically.[15]

The unfolding transition is often cooperative, with dramatic changes occurring over a very narrow range of temperature or within a narrow concentration band of denaturant. Note that thermal unfolding is usually performed at low pH to minimize disulphide interchange in the unfolded protein.[16] Measurement of a protein's T_m (the transition midpoint of thermal unfolding, sometimes described colloquially as the 'melting' or unfolding temperature) is usually carried out spectrophotometrically at a single diagnostic wavelength, using a thermal programmer which gradually raises the cell temperature over a set range. T_m is a robust parameter, which is purely empirical and involves no underlying assumptions. Empirical stability parameters such as the T_m or the denaturant concentration midpoint, can easily be estimated from experimental data. These measurements will suffice to rank a series of mutant or variant proteins in order of stability, even where a two-state model (see next paragraph) does not apply.[16,17]

Calculation of Gibbs energies may be performed on thermal data of this sort or on unfolding studies using urea or guanidine hydrochloride, but requires care. Thermodynamic analysis of unfolding data is valid only if the observed transition is two-state; Shirley has described how one may check that this is so.[18] A two-state transition involves no intermediates and occurs in a single step without any plateau or shoulder. It should yield identical patterns for different observational methods (e.g. circular dichroism, differential scanning calorimetry or fluorescence measurements) and enthalpy values determined calorimetrically should agree with those calculated by the van't Hoff equation ($d\ln K/dT = \Delta H/RT^2$, where K is the equilibrium constant, T is the temperature in Kelvins, ΔH is the heat of reaction and R is the gas constant).[18] Baldwin and Eisenberg have also outlined criteria for a two-state process.[16] Gibbs (free) energies of unfolding must be calculated from a careful and rigorous analysis of thermal or urea/guanidine hydrochloride denaturation curves. These energies are termed $\Delta G(25°C)$ and $\Delta G(H_2O)$, respectively. A typical curve will be sigmoidal in shape with linear portions at the extremes and a sharp, deep transition in the middle (see Fig. 1.4). The linear extremes of the plot (i.e. points at the lowest and highest temperatures or denaturant concentrations) must be determined accurately (by inclusion of many experimental points) since they need to be extrapolated into the transition region (Fig. 1.4). The purpose of this is to calculate the fraction of native protein (F_N) at each point. This is given by the equation $F_N = (y - y_U)/(y_N - y_U)$ where y is the experimental value at a particular point and y_N and y_U are the values for the native and unfolded species extrapolated to the same position in the transition region. Using F_N, one can then estimate the Gibbs energy difference from the equations $K = F_N/(1 - F_N)$ and $\Delta G° = -R.T.\ln K$, where K is the equilibrium constant, R the gas constant (in units of $J.mol^{-1}.K^{-1}$) and

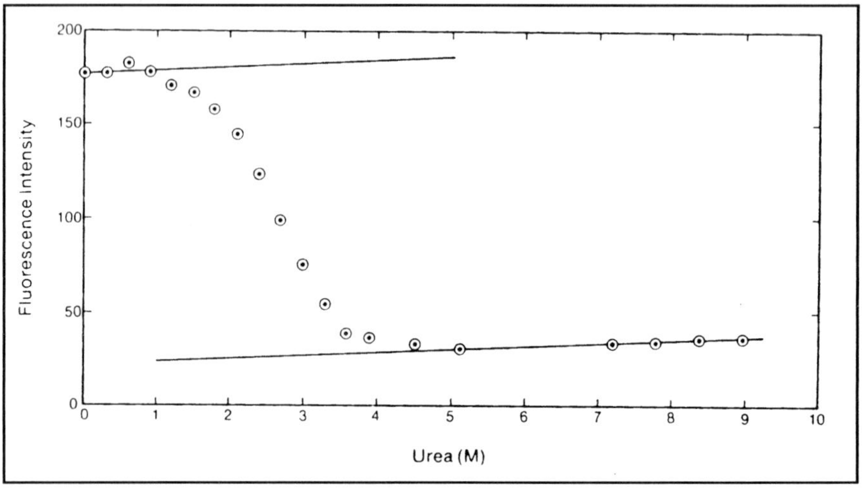

Fig. 1.4. Unfolding curve for ribonuclease a. Reprinted with permission from: Pace CN. Meths Enzymol 1986; 131:266-280. © 1986 Academic Press Inc.

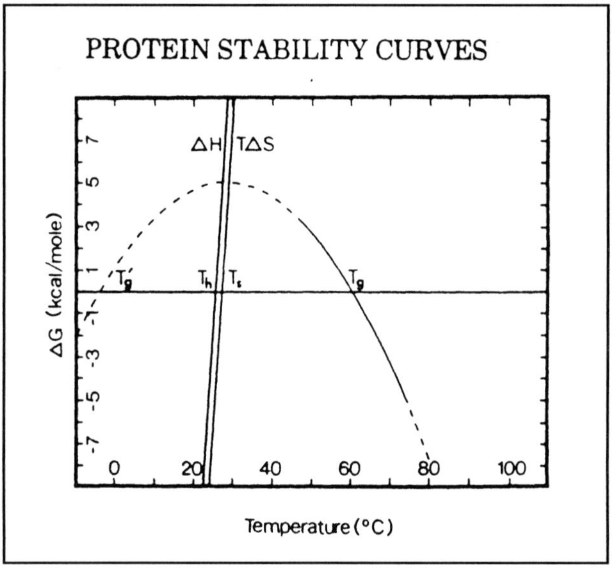

Fig. 1.5. A stability curve for a protein. The solid line represents that portion of the curve used in a van't Hoff analysis of the transition while the dashed line is extrapolated. T_g', the low-temperature melting point; T_h, the zero-enthalpy temperature; T_s, the temperature of maximum stability (where entropy is zero); T_g, the high-temperature melting point. Reprinted with permission from: Becktel WJ, Schellman JA. Biopolymers 1987; 26:1859-1877. © 1987 John Wiley & Sons.

T the temperature in Kelvins.[16] Refs. 19-21 contain fuller discussions of Gibbs energy measurement. Calorimetric techniques such as differential scanning calorimetry allow direct calculation of thermodynamic parameters by measuring changes in specific heat capacity.[22]

There is also a temperature where the entropy is zero. This is the temperature of maximum stability, T_S, which may be obtained in suitable cases from a plot of ΔG against temperature (a stability curve, Fig. 1.5). Three postulates must be satisfied for a valid stability curve, namely:

1) the denaturation transition is a two-state process,
2) the protein is stable at *some* temperature, i.e., the Gibbs energy is positive over some range of temperature, and
3) the change in heat capacity, ΔC_p, on unfolding of the protein has a constant positive value (see next paragraph, however).

Experimental points define a segment of a bell-shaped theoretical curve. Variables or events such as pH, chemical modification, ligand binding and mutations can affect the stability curve's position and shape. Becktel and Schellman discuss the concepts underlying stability curves in ref. 23.

Certain proteins are more stable at room temperature than in the refrigerator and are said to be *cold labile*. This cold denaturation has been well characterized for myoglobin and a few other proteins.[23-25] It is a property of the protein itself and is distinct from freezing inactivation (see chapter 9). This phenomenon arises from the fact that it is thermodynamically possible for a protein to unfold at low as well as at high temperatures (see ref. 26 for a summary of the notable features of cold denaturation). It is possible to study cold denaturation directly using certain experimental tricks,[26] including the use of salts, chaotropes or pH to bring the unfolding temperature within an experimentally-accessible range, organic cosolvents to depress the medium's freezing point below the unfolding transition temperature, use of partially destabilized mutants or undercooling (see chapter 9 for a note on the latter). Franks advises direct study of cold denaturation at low temperatures and cautions against reliance on extrapolated heat denaturation data, since there is evidence that ΔC_p *does* vary with temperature.[26] Cold denaturation is completely reversible[26] but it may cause unsuspecting scientists some unpleasant shocks!

Kinetic Stability

Kinetic stability is distinct from (and need not correspond with) *thermodynamic* stability. It involves measuring the persistence of catalytic (or other biological) activity with time under adverse conditions of temperature, pH, solvents, salt concentration and so on (or, to put it another way, the progressive loss of function). It can be represented by the scheme

$$N \xrightarrow{k_{in}} I$$

where N is the native, functional protein, I is an irreversibly inactivated form and k_{in} is the rate constant for the inactivation process. The equation $V_{in} = -d[N]/dt = k_{in}[N]$ describes the process mathematically, where V_{in} is the experimentally observed rate of disappearance of the native form.[10] This is usually equated with the molecule's long-term stability.[10] Experimentally, determination of kinetic stability is fairly simple. The protein of interest is incubated at a 'suitable' elevated temperature, i.e., one where activity loss is neither too rapid nor inconveniently slow. Samples are withdrawn at intervals, cooled to a constant lower temperature

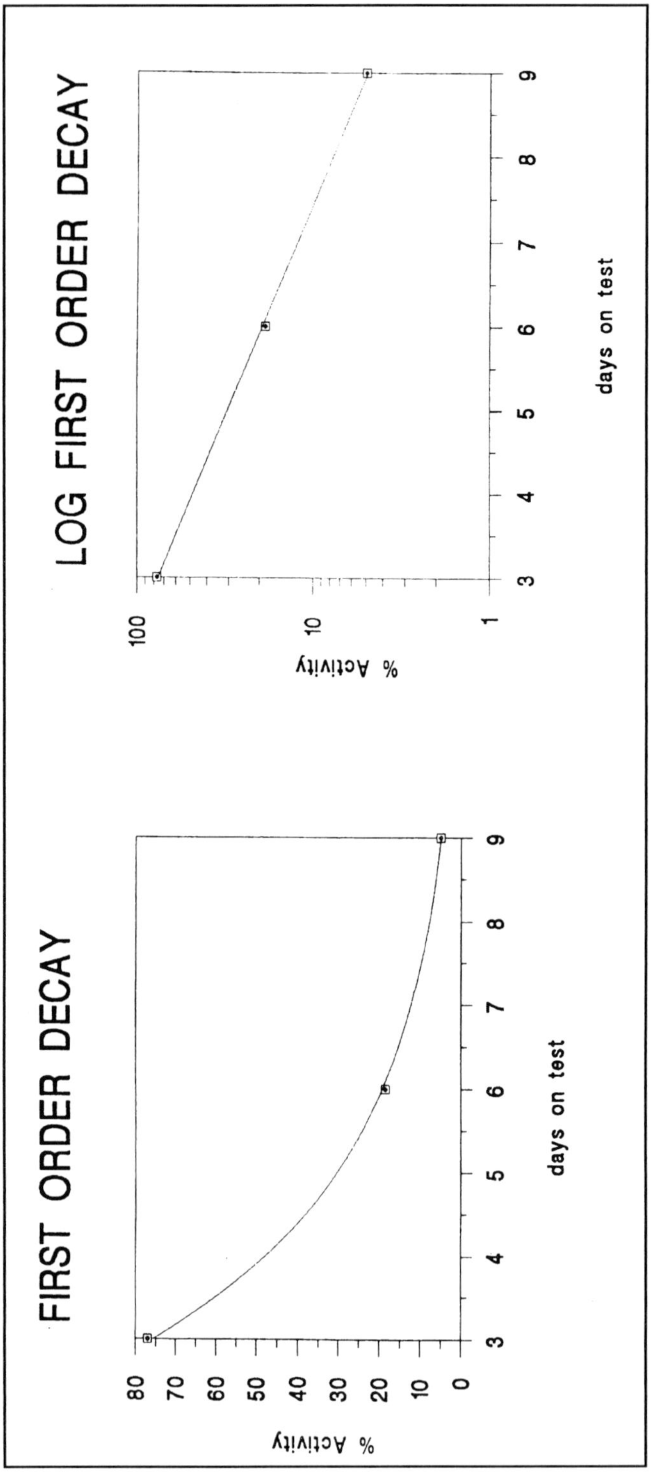

Fig. 1.6. Exponential decay of enzyme activity at elevated temperature.

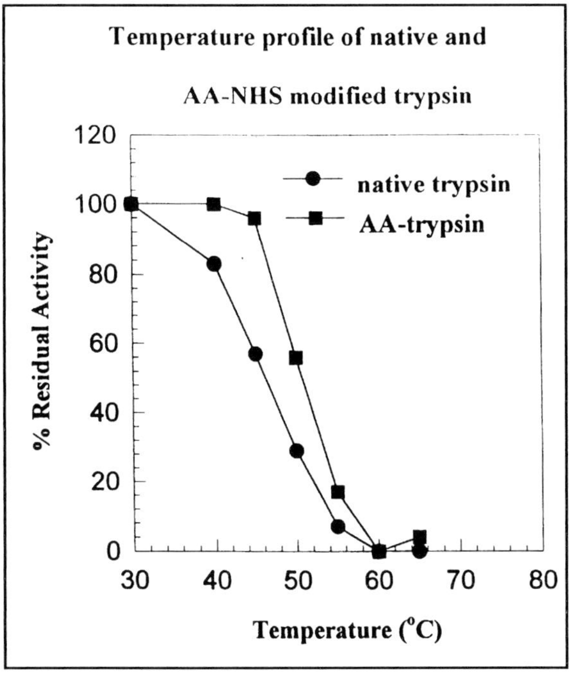

Fig. 1.7. Temperature profile for native and acetylated beef trypsin. Reprinted from: J Biotechnol 49: Murphy A, Ó Fágáin C. Stability characteristics of chemically-modified soluble trypsin, 163-171, 1996, with kind permission of Elsevier Science - NL. © 1996 Elsevier Science - NL, Sara Burgerhartstraat 25, 1055 KV Amsterdam, The Netherlands.

and assayed for activity under optimal conditions. If activity at time zero is defined as 100%, a plot of percent activity remaining against time may be used to estimate and compare stabilities. The time required to reach 50% of initial activity can be estimated by inspection of the plot. If activity loss fits a single exponential function (i.e. if a plot of ln(activity) versus time is linear) then the first-order rate constant k may be used to calculate a true half-life (see Fig. 1.6). (Recall that the equation for a first-order [single] exponential decay is $A_t = A_0\exp[-kt]$, where A_0 and A_t represent the activities at times zero and t, respectively, and k is the first-order rate constant. Half-life is defined as $0.693/k$.) Often the activity loss will be first-order, although more complex inactivation patterns are well documented.[27] Note, however, that an apparently unimolecular first-order time course of inactivation may mask a more complex set of inactivating molecular events.[27] k-Values can also be informative when correlated with other data such as the number of attachment sites to a solid phase (immobilized enzyme) or changes in protein hydrophilicity.[10]

A *temperature profile* can be determined from an experiment in which the enzyme or protein is incubated for a fixed, short period (typically 10 or 15 minutes) at successively increasing temperatures, following which the samples' residual activities are measured. Often a sigmoidal curve like that depicted in Figure 1.7 results, and it is possible to determine a T_{50} temperature, analogous to T_m, where activity drops to 50% of the initial value over the time period of incubation. The T_{50} and T_m values may well be equivalent in the case of a simple monomeric protein, but one should note that they need not agree, especially with oligomeric proteins or proteins containing prosthetic groups.

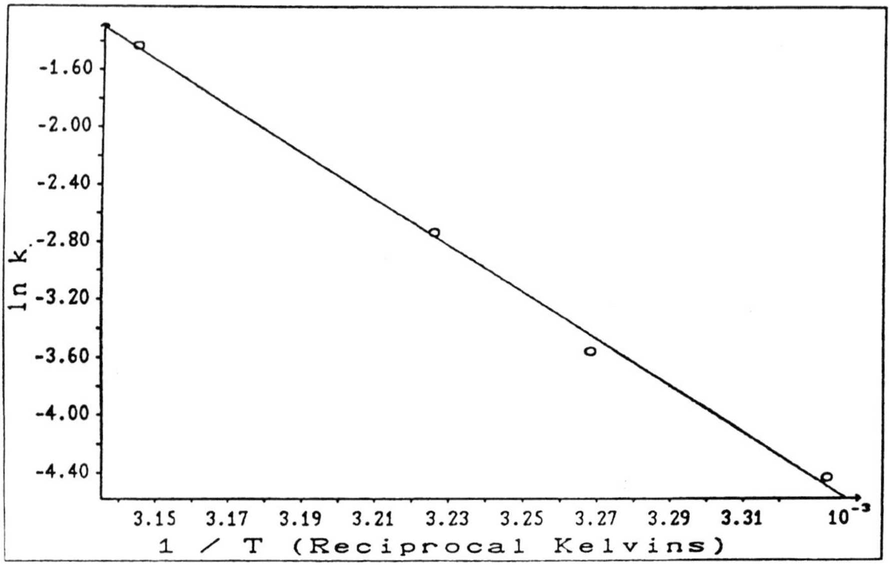

Fig. 1.8. Arrhenius plot. Arrhenius plot of ln (first-order rate constant) against reciprocal temperature (Kelvins, 1/T). Reprinted with permission from: Ó Fágáin C. Am Biotechnol Lab 1991; 9(9):32. © 1991 International Scientific Communications Inc.

Accelerated Degradation Testing

Kinetic stability is usually measured at elevated temperatures[10] but the inactivating event(s) at high temperatures may not mirror that/those at the much lower temperatures used for storage. It is not feasible, however, to monitor stability in real time at the actual storage temperature: the experiment would take too long. Inaccuracy may result over shorter intervals, since only minimal losses, virtually indistinguishable from the starting activity, would be apparent. Fortunately, there is a methodology which can overcome these difficulties in many cases, namely accelerated degradation (or accelerated storage) testing. This makes use of the first-order rate constant of inactivation, k, described above. The technique involves placing samples of the protein of interest at various elevated temperatures and monitoring the decline in activity over time against material held at a low reference temperature, at which one assumes no activity loss takes place. One then uses the Arrhenius equation to predict 'shelf lives' at temperatures of interest. The Arrhenius equation is conveniently expressed in logarithmic form as $\ln k = -E_a/R.T + \ln A$, where k is the first-order rate constant of activity decay (with units of $time^{-1}$), E_a is the activation energy (of the molecular event leading to inactivation), R is the gas constant (in units of $J.mol^{-1}.K^{-1}$), T is the temperature in Kelvins and A represents the initial activity. This log form of the Arrhenius equation yields a straight line plot of $\ln k$ against 1/T with slope $-E_a/R$ (ref. 28; see Fig. 1.8). The Arrhenius equation is empirical but a similar equation results from the applications of Eyring's transition state theory (see ref. 28 for a fuller discussion of this point). Extrapolation of this plot can give the rate constant (and hence the useful life) at a particular temperature. The ability to predict the duration of biological activity can be very useful, especially in the pharmaceutical industry. Accelerated storage testing has been used as a practical means of quality assurance for biological standards[29] and has featured in some investigations of protein thermostabilities.[30,31] Kirkwood has given a number of practical hints on the conduct of accelerated storage determinations[32] while Franks has outlined some pitfalls to be avoided when using this methodology.[33,34] Errors may arise in situations where:

1) activity loss is due to two (or more) reactions with different activation energies, occurring either simultaneously or consecutively,
2) a glass transition (see chapter 9) occurs within the product over the measured temperature range, and
3) a phase transition occurs within a solid product.[34]

One should also note that Šikšnis and colleagues observed an entirely novel phenomenon while studying stabilized, chemically modified forms of alpha-chymotrypsin. Unlike native alpha-chymotrypsin and most of its chemical derivatives, some 'hydrophilized' forms exhibited a 'zig-zag' pattern in the Arrhenius plots of their irreversible inactivation rate constants which had never been reported before. The inactivation rate increased with temperature between 45-55°C, giving rise to the 'normal' Arrhenius plot with a negative slope. Between 55-65°C, the inactivation rate constant *decreased* with increasing temperature, giving rise to a *positive* slope in that portion of the Arrhenius plot. Finally, above 75°C, the inactivation rate constant again increased with temperature, leading to a negative slope in this third segment of the Arrhenius plot. The researchers explained these unprecedented findings by postulating a reversible conformational transition between 50-60°C in their preparations of chemically modified alpha-chymotrypsin. Fluorescence evidence supported this postulate and the authors went on to develop a model explaining this highly unusual behavior (see ref. 35 and chapter 6).

Setting Up an Accelerated Degradation Test

An accelerated storage test must use the minimum amount of the (often precious) test protein that is compatible with achievement of precise and accurate results. Since each experiment will occupy a period of weeks, it is important that every test yields a meaningful outcome. One must therefore take great care in setting up an accelerated degradation test run.

It is essential to prevent microbial contamination of test samples (see chapter 9), especially since one assumes that no (or very little) sample degradation takes place at the low reference temperature during the period of the experiment. Place test samples at a series of elevated temperatures (e.g. 48°, 45°, 42°, 37°, 33°, 30°C) and at a suitable, lower reference temperature, say, 4°C. It can be a good idea to place a maximum/minimum thermometer inside the refrigerator, freezer or incubator(s) close to the containers of interest in order to record any variations of temperature which may occur over an extended period. (First check that the thermometer will withstand the low or high temperatures!) Ensure that the reference temperature will not cause freezing of liquid samples: a liquid-solid phase change will introduce a further variable and may also lead to freezing damage (see chapter 9). Remove samples at intervals from the various incubation conditions, bring them to the same temperature and assay for activity under standardized, optimal conditions. Ensure that stock solutions used in assays are carefully prepared and standardized: it is likely that different batches of assay solutions will be used over the period of the accelerated storage test, so variations must be minimized. One must be scrupulous about procedural details of the assay and the quality and performance of the instruments used.

Kirkwood has made some practical recommendations for successful accelerated storage testing[32] which are summarized here. Use at least 3 elevated temperatures plus a low reference temperature. Set up 10 or more samples at each of these temperatures. The testing schedule set out below avoids waste of material due to testing at inappropriate intervals. It also allows checking of the order of reaction. An acceptable result is very likely if enough material is placed on test at the beginning.

1. At intervals, test samples stored at the highest temperatures against the low temperature reference samples. Ignore intermediate temperatures until a loss of 25% or greater (against the low temperature reference sample) has occurred at the highest temperature. Now test a second sample to confirm the first result. If the results disagree, continue this stage 1 testing. If they agree, move on to stage 2.
2. Test the next two highest temperatures against a reference sample. Fit all available data to the Arrhenius equation. This procedure will not give a final precise result but it will facilitate progress to stage 3.
3. Use the approximate result from stage 2 to ascertain further periods of storage at all temperatures such that measurable activity losses will have occurred at the intermediate temperatures. Assay samples from all available temperatures and check all available data for their ability to fit the Arrhenius equation. This should result in a reasonable estimate of the low temperature degradation rate, especially if multiple assays have been performed. If too little degradation has occurred to yield a precise result, one can repeat stage 3 after a longer time.[32]

Analysis of Accelerated Degradation Results

Clearly, 'raw' experimental data must undergo a number of transformations for use in the Arrhenius plot, notably conversion to natural log or reciprocal forms. Error relationships are often significantly affected by these sorts of transformations. Use a computer for all statistical fitting to minimize such errors. Use of good quality replicate results is very important.

The activity decay must be first-order at all of the temperatures used in the Arrhenius plot. Verify this by fitting the time course of activity loss at each test temperature to a first-order exponential decay. Assess goodness of fit by inspection of the graphic fit and of parameters such as standard errors or chi-square values. First-order exponentials yield straight line plots when transformed into semi-logarithmic form, unlike higher order functions (i.e. a plot of ln[Activity] against time is linear for a first-order decay). Deviations from a first-order function are more likely to occur at higher temperatures.[29] Amorphous solid preparations will follow Arrhenius kinetics provided they remain in the glassy state (chapter 9). However, if any of the elevated temperatures used exceeds the glass transition temperature (T_g'), the product will no longer obey the Arrhenius equation and will likely deteriorate at a much faster rate. See ref. 33 for a description of this phenomenon and of an alternative kinetic scheme. Other situations in which deviations from Arrhenius kinetics may occur are outlined in ref. 34.

Using the k-values determined at different temperatures, plot the Arrhenius graph [ln$k$ is the ordinate (Y-axis), 1/T the abscissa (X-axis)]. A linear plot of negative slope results. Extrapolate the line to a temperature of interest (e.g. 0°, 4° or 25°C; these are, respectively, 273, 277 and 298 Kelvins) and estimate the value of k at this temperature. It is easy to estimate a true half life at this temperature by using

Table 1.1. Principal indices of protein stability

Parameter	Measures	Requirements	How estimated	Ref.
% Activity	Residual activity at time *t*	None	$Activity_{time\ t}/Activity_{time\ 0} \times 100\%$	(a)
Half life	Time to reach 50% activity.	First-order decay	$0.693/k$ (k = rate constant)	(a)
Accelerated degradation	Lifetime at *T*°C at all temperatures used	First-order decay	Extrapolation of plot ln$k$ v $1/T$ (Kelvin)	(b-d)
T_m	Melting temperature	None	Temperature where unfolding = 50%	(e)
$\Delta G_{(25°C)}$	Conformational stability	Two-state unfolding	Thermal denaturation curve	(f)
$\Delta G_{(H_2O)}$	Conformational stability	Two-state unfolding	Denaturant unfolding curve	(f)
T_S	Temperature of maximum stability	Two-state unfolding	Stability curve	(g)

(a) Mozhaev VV. Mechanism-based strategies for protein thermostabilization. Trends Biotechnol 1993; 11:88-95.
(b) Jerne NK, Perry WLM. The stability of biological standards. Bull Wld Hlth Org 1956; 14:167-182.
(c) Kirkwood TBL. Design and analysis of accelerated degradation tests for the stability of biological standards III. Principles of design. J Biol Stand 1984; 12: 215-224.
(d) Franks F. Accelerated stability testing of bioproducts: attractions and pitfalls. Trends Biotechnol 1994; 12:114-117.
(e) Baldwin RL, Eisenberg DE. Protein stability. In: Oxender DL, Fox CF., eds. Protein Engineering. NY: Alan R Liss, 1987:127-148.
(f) Pace CN. Measuring and increasing protein stability. Trends Biotechnol 1990; 8:93-98.
(g) Becktel WJ, Schellman JA. Protein stability curves. Biopolymers 1987; 26:1859-1877.

the equation $t_{1/2} = 0.693/k$. Specially designed programs for the analysis of accelerated storage data are available. Below is a brief description, for information only, of two programs for analysis of accelerated degradation data. Their underlying statistical methodology is beyond the scope of this chapter and no endorsement or judgment of their merits or claims is made or implied. Each user must determine the suitability of any package for his/her purposes.

The DEGTEST program carries out a three-stage analysis of data. First, the % activities at each temperature (relative to the low temperature reference sample) are used to give estimates of relative degradation rates, assuming first-order decay. Second, these rates are fitted to the Arrhenius equation and the goodness of fit is ascertained. Third, the decay rate at the proposed storage temperature is predicted from the Arrhenius fit and the statistical precision of this value is calculated. The program can fit to the Arrhenius equation or to the more rigorous Eyring equation.[36] In practice, however, little difference is observed between the two equations.[37] The POTENCYLOSS program is based on a linear model for stability prediction which uses a double logarithmic primary plot together with an altered secondary plot corresponding to the Arrhenius plot. The program can distinguish between zero- and first-order data and gives loss rate and shelf life parameters at a desired designated temperature. In addition, one can obtain estimates of activity values at any temperature or future time period.[38]

Another use of the Arrhenius plot in protein stability measurement has been described by Voordouw et al.[39] These workers used Arrhenius methodology to determine the contribution of structural metal ions (specifically, calcium in proteolytic enzymes) to the overall observed kinetic stability. They reasoned that the total kinetic thermostability of a calcium-containing protease contains contributions from the polypeptide chain and from the 'structural' calcium ion(s) and that it should be possible to separate these two components. This separation can be accomplished by expressing the thermal stability as the total free energy of activation in the thermal denaturation reaction(s). A theoretical framework for this methodology was formulated. Values for the denaturation rate constant, k, at zero and at saturating calcium concentrations are obtained from extrapolated Arrhenius plots. (These must show good linearity characteristics if the procedure is to be valid, and the extrapolation must not extend over too wide a temperature range). Experimentally, one determines Arrhenius plots at high and zero concentrations of free calcium ion. (An excess of EDTA can be used to remove calcium from the system.) Analysis of six calcium-containing proteases at their respective optimal pH values for stability revealed notable differences. (Only the forward denaturation was studied, since autolysis would likely have rendered the denaturation irreversible.) Among the three serine proteases, thermomycolase and subtilisins Carlsberg and BPN' (which each bind a single calcium ion with high affinity), thermomycolase displayed the highest stability at 70°C. This was ascribed mainly to the intrinsic contribution of the protein chain. Two metalloendopeptidases (thermolysin and neutral protease A) were also examined. Thermolysin was the more stable and this seems to arise from a calcium ion effect. Remarkably, the intrinsic stability of the thermolysin polypeptide is probably of the same order as that of neutral protease A and is smaller than that of the three serine enzymes studied. It seems that thermolysin's high thermal stability arises from the high-affinity binding ($K \sim 6 \times 10^7\ M^{-1}$ at 80°C) of a single calcium ion to one of the four calcium binding sites.

Other indices of enzyme stability include the decimal reduction time D, the time required for a 10-fold decrease in initial activity ($D = \ln(10)/k$) and the z-value, the temperature increase required for a 10-fold reduction of the D-value ($D = D_{ref}.10^{(T-Tref)/z}$ where D and D_{ref} stand for the decimal reduction times at temperatures T and the reference temperature T_{ref} (°C), respectively).[40,41]

To conclude, there are many indices of protein stability. The most prominent of these are summarized in Table 1.1. One should choose whichever is most suitable for one's purposes and avoid inappropriate comparisons. Practical limitations of the method used should be thoroughly understood and all data analysis should be statistically rigorous. The loss or decrease of the protein's biological or functional activity will be the main and most important index of deterioration. Often, however, the degree or time course of activity loss will not give any indication of the underlying molecular cause (although aggregation may be readily visible). Ref. 10 provides a useful table of methods to identify the molecular changes leading to inactivation of the target protein.

References

1. Volkin DB, Klibanov AM. Minimizing protein inactivation. In: Creighton TE, ed. Protein Function: a Practical Approach. Oxford, IRL Press, 1989:1-24.
2. Voet D, Voet JG. Biochemistry. 2nd edn. New York: J Wiley and Sons, 1995.
3. Creighton TE. Proteins: Structure and Molecular Properties. 2nd edn. San Francisco: WH Freeman, 1990.
4. Mozhaev VV, Martinek K. Inactivation and reactivation of enzymes. Enzyme Microb Technol 1982; 4:299-309.
5. Parsell DA, Sauer RT. The structural stability of a protein is an important determinant of its proteolytic susceptibility in *Escherichia coli.* J Biol Chem 1989; 264: 7590-7595.
6. Gething M-J, Sambrook J. Protein folding in the cell. Nature 1992; 355:33-45.
7. Martin J, Mayhew M, Langer T et al. The reaction cycle of GroEL and GroES in chaperonin-assisted protein folding. Nature 1993; 366:228-242.
8. Martin J, Geromanos S, Tempst P et al. Identification of nucleotide-binding regions in the chaperonin proteins GroEL and GroES. Nature 1993; 366:279-282.
9. Parsell DA, Kowal AS, Singer MA et al. Protein disaggregation mediated by heat-shock protein Hsp104. Nature 1994; 372:475-478.
10. Mozhaev VV. Mechanism-based strategies for protein thermostabilization. Trends Biotechnol 1993; 11:88-95.
11. Zale SE, Klibanov AM. On the role of reversible denaturation (unfolding) in the irreversible thermal inactivation of enzymes. Biotechnol Bioeng 1983; 25: 2221-2230.
12. Schmid FX. Spectral methods of characterizing protein conformation and conformational changes. In: Creighton, TE, ed. Protein Structure: a Practical Approach. Oxford: IRL Press, 1989:251-285.
13. Goldenberg DP. Analysis of protein conformation by gel electrophoresis. In: Creighton, TE, ed. Protein Structure: a Practical Approach. Oxford: IRL Press, 1989:239-250.
14. Hilser VJ, Worosila GD, Freire E. Analysis of thermally induced protein folding/ unfolding transitions using free solution capillary electrophoresis. Anal Biochem 1993; 208:125-131.
15. Arakawa T, Hung L, Pan V et al. Analysis of the heat-induced denaturation of proteins using temperature gradient gel electrophoresis. Anal Biochem 1993; 208:255-259.
16. Baldwin RL, Eisenberg DE. Protein stability. In: Oxender DL, Fox CF, eds. Protein Engineering. NY: Alan R Liss, 1987:127-148.
17. Tombs MP. Stability of enzymes. J Appl Biochem 1985; 7:3-24.

18. Shirley BA. Protein conformational stability estimated from urea, guanidine hydrochloride and thermal denaturation curves. In: Ahern TJ, Manning MC, eds. Stability of Protein Pharmaceuticals, Part A: Chemical and Physical Pathways of Protein Degradation. NY: Plenum Press, 1992:167-194.
19. Pace CN. Measuring and increasing protein stability. Trends Biotechnol 1990; 8:93-98.
20. Pace CN, Shirley BA, Thomson JA. Measuring the conformational stability of a protein. In: Creighton, TE, ed. Protein Structure: a Practical Approach. Oxford: IRL Press, 1989:311-330.
21. Pace CN. Determination and analysis of urea and guanidine hydrochloride denaturation curves. Meths Enzymol 1986; 131:266-280.
22. Volkin DB, Middaugh CR. The effect of temperature on protein structure. In: Ahern TJ, Manning MC, eds. Stability of Protein Pharmaceuticals, Part A: Chemical and Physical Pathways of Protein Degradation. NY: Plenum Press, 1992: 215-247.
23. Becktel WJ, Schellman JA. Protein stability curves. Biopolymers 1987; 26:1859-1877.
24. Privalov PL. Cold denaturation of proteins. CRC Crit Rev Biochem Mol Biol 1990; 25:281-305.
25. Franks F, Hatley RHM. Stability of proteins at subzero temperatures: thermodynamics and some ecological consequences. Pure Appl Chem 1991; 63: 1367-1380.
26. Franks F. Conformational stability of proteins. In: Franks F, ed. Protein Biotechnology: Isolation, Characterization and Stabilization. Totowa, NJ: Humana Press, 1993:395-436.
27. Sadana A. Enzyme deactivation. Biotechnol Adv 1988; 6:349-446.
28. Eisenberg DS, Crothers I, Donald M. Physical Chemistry. Menlo Park, CA: Benjamin/Cummings 1979:239-245.
29. Jerne NK, Perry WLM. The stability of biological standards. Bull Wld Hlth Org 1956; 14:167-182.
30. Malcolm BA, Wilson KP, Matthews BW et al. Ancestral lysozymes reconstructed, neutrality tested, and thermostability linked to hydrocarbon packing. Nature 1990; 345:86-89.
31. Ó Fágáin C, O'Kennedy R, Kilty C. Stability of alanine aminotransferase is enhanced by chemical modification. Enzyme Microb Technol 1991; 13:234-239.
32. Kirkwood TBL. Design and analysis of accelerated degradation tests for the stability of biological standards III. Principles of design. J Biol Stand 1984; 12:215-224.
33. Franks F, Hatley RHM, Mathias SF. Materials science and the production of shelf-stable biologicals. Pharm Technol Int 1991; 3:24-34.
34. Franks F. Accelerated stability testing of bioproducts: attractions and pitfalls. Trends Biotechnol 1994; 12:114-117.
35. Siksnis V, Mozhaev VV, Galkantaite N et al. Unusual (zig-zag) temperature dependence of the rate constant for irreversible inactivation of hydrophilized enzymes. Collect Czech Chem Commun 1990; 55:1366-1371.
36. Kirkwood TBL, Tydeman MS. Design and analysis of accelerated degradation tests for the stability of biological standards II. A flexible computer program for data analysis. J Biol Stand 1984; 12:207-214.
37. Tydeman MS, Kirkwood TBL. Design and analysis of accelerated degradation tests for the stability of biological standards I. Properties of maximum likelihood estimators. J Biol Stand 1984; 12:195-206.
38. Nash RA. A new linear model for stability prediction. Drug Dev Ind Pharm 1987; 13:487-499.

39. Voordouw G, Milo C, Roche RS. Role of bound calcium ions in thermostable, proteolytic enzymes. Separation of intrinsic and calcium ion contributions to the kinetic thermal stability. Biochemistry 1976; 15:3716-3724.
40. Hendrickx M, Saraiva J, Lyssens J et al. The influence of water activity on thermal stability of horseradish peroxidase. Int J Food Sci Technol 1992; 27:33-40.
41. Weng Z, Hendrickx M, Maesmans G et al. Thermostability of soluble and immobilized horseradish peroxidase. J Food Sci 1991; 56:574-578.

CHAPTER 2

Studies on Denaturation, Inactivation and Stabilizing Interactions

Introduction

The crucial importance of a protein's tertiary structure, i.e., its molecular shape, has been remarked upon in chapter 1. Tertiary structure arises from interactions between the side chains (or R-groups) of the covalently-linked amino acids making up the polypeptide. A complex and exact pattern of chain folding is superimposed on the ordered regions that comprise the secondary structure. It is the tertiary structure that orientates the critical residues and side chains into the correct geometrical relationship to permit function. This is not to state that a protein molecule is completely rigid, however. There is abundant evidence that enzymes and proteins undergo slight but significant changes in shape on binding substrates or modulators. The concept of 'induced fit' is well established in enzymology and changing interactions between the four subunits of hemoglobin are known to underlie that molecule's positive cooperativity with respect to oxygen binding. It may help to think of a properly folded protein as a pasta shape cooked 'al dente': it is flexible within limits while retaining a well-defined, recognizable shape. On the other hand, an unfolded or denatured protein (chapter 1) approximates more to a strand of fully-cooked spaghetti in that it is completely flexible and lacks any defined shape. What causes the pasta shape to unravel to resemble a spaghetti strand? What happens to the unravelled spaghetti? This chapter examines denaturation or unfolding of a protein and the nature of subsequent molecular changes leading to irreversible inactivation. It also surveys explorations of stabilizing interactions in proteins and considers the contribution of the carbohydrate portions of glycoproteins to stability.

Denaturation Studies

Armed with the parameters described in chapter 1 for measurement of protein stability, researchers have been able to study the loss of protein function or structure in a rational fashion. Understanding the causes of activity loss can help in the formulation of stabilizing strategies, since one will know what changes must be prevented. Many enzyme deactivations have been characterized in detail. A first-

Stabilizing Protein Function, by Ciarán Ó Fágáin.
© Springer - Verlag and Landes Bioscience 1997.

order exponential process (i.e. a unimolecular decay) describes many deactivations, since two-state transitions (Native $\leftrightarrow$ Unfolded; see chapter 1) are often observed in reality. More complex phenomena do occur, however, but even some of these have been successfully modelled. Sadana reviewed first order deactivation processes with differing parameters.[1] His article includes a survey of complex enzyme deactivations, especially non-series ones. Oligomeric proteins are more likely to undergo complex deactivation. Many monomeric proteins show two-state unfolding but there are exceptions. The single subunit protease zymogen, pepsinogen, undergoes a transition that is not two-state.[2]

One can observe denaturation of most proteins at high temperatures. The conformational stability of proteins is due to the (quite small) net difference between a very large number of weak stabilizing interactions and the nearly-equally large conformational entropy. This net free (Gibbs) energy of stabilization (typically 40 $kJ.mol^{-1}$) is equivalent to that of a small number of interactions. Only a very few further interactions is sufficient to explain the greater stabilities of very stable or extremophilic proteins: a single interaction may contribute up to 25 $kJ.mol^{-1}$ (ref. 3). Some proteins, however, will denature at low temperatures[4] and this has been well described for metmyoglobin.[5] Dill and colleagues included this work in their development of a statistical thermodynamic theory to predict the temperature dependence of globular protein stability.[6] Privalov has reviewed the topic of cold denaturation of proteins in ref. 7.

There are examples where the unfolding temperature of a protein of interest decreases greatly in the presence of chaotropic agents.[8,9] Findings of this sort show that different denaturing influences (whether temperature, pH or chaotropic agents) act in an equivalent fashion by encouraging unfolding of the three-dimensional protein structure. Where inactivation takes place, some covalent change or alteration in the degree of association occurs instead of, together with, or in addition to, the unfolding phenomenon. Activity will be lost when the unfolding disrupts the integrity of the molecule's active or functional site(s). In summary, loss of a protein's biological activity can occur by either conformational or covalent processes. The conformational processes are often reversible and correspond to Mozhaev and Martinek's definition of 'denaturation' in ref. 8.

Zale and Klibanov have described a theoretical and practical investigation of the contribution of reversible unfolding (denaturation) to the overall irreversible thermal inactivation of ribonuclease from beef pancreas. No visible precipitates formed and no reduction in protein concentration occurred. The inactivation time course followed strict first order (unimolecular) kinetics. Aggregation phenomena are oligo- or polymolecular and depend on protein concentration,[10] so aggregation cannot be the cause of the activity loss. Also, plots of % initial activity remaining versus time were independent of protein concentration over a wide range. A model involving three enzyme states (N, U and I) was proposed. N signifies native enzyme which is in equilibrium with the reversibly unfolded, denatured form U. (Zale and Klibanov actually called the denatured/unfolded protein molecule 'D' but the designation 'U' is retained here for consistency with chapter 1.) K is the equilibrium constant for the N $\leftrightarrow$ U transition. I is the irreversibly inactivated form and k is the rate constant for its formation from U (see Fig. 1.3). Using this model a simple equation was derived relating the experimentally determined inactivation rate constant k_{obs} and the true monomolecular rate constant k: $k_{obs} = k/(1+K)$. Reversible unfolding always occurred before the irreversible inactivation. At lower temperatures, resistance both to reversible denaturation and to irreversible inactivation correlated well; i.e. the more the enzyme resists reversible unfolding, the more stable it will be against irreversible activity loss. This relationship breaks down above a certain temperature, however. Therefore, any stabilization strategy based on the prevention of unfolding will yield an apparently stabilized enzyme only if inactivation rate measurements are carried out below the critical temperature. This will vary with the protein under study. These findings also have implications for the extrapolations of high temperature data used in Arrhenius plots.[11]

Air-liquid interfaces can also have important destructive effects on proteins, especially under conditions of rapid mixing or agitation, as has been demonstrated for a variety of proteins including lipase. Surfactants which preferentially adsorb to the air-water interface can prevent inactivation, providing further evidence for this interfacial effect.[12,13] Hoare et al used yeast alcohol dehydrogenase as a model protein to observe the effects of various types of process plant. Both laboratory and pilot scale apparatus were used.[12] Provided that no air entrainment took place, pumps had scarcely any adverse effect on activity until very high recycle numbers (>1,000) were attained. Within these limits, the pump type (centrifugal, peristaltic, gear or mono screw) had virtually no effect. Ultrafiltration membranes were likewise without adverse effect except for some surface adsorption to a fresh membrane. Damage did occur to yeast alcohol dehydrogenase during passage through polymeric microfiltration membranes, perhaps due to combined shear and interfacial effects.[12,14]

Deleterious Chemical Reactions in Proteins

The purely chemical reactions are few and well defined. Deamidation of glutamine and asparagine can occur at neutral to alkaline pH values while peptide bonds involving aspartic acid undergo cleavage under acidic conditions. Cysteine is prone to oxidation, as are tryptophan and methionine. Alkaline conditions lead to reduction of disulfide bonds and this is often followed by beta-elimination or thiol-disulphide exchange reactions.[15] Where reducing sugars are present with free protein amino groups (N-termini or lysine residues), there may be destructive glycation of the amino functions by the reactive aldehyde or keto groups of the sugar (the Maillard reaction).[16] Most of these reactions are discussed in greater detail in the paragraphs below. Elevated temperatures favor all of these reactions but it is important to note that aggregation and deleterious chemical reactions can occur at moderate temperatures also. Virtually complete aggregation of lyophilized bovine serum albumin occurred over 24 hours at 37°C following addition of just 3 microliters of physiological saline. The degree of aggregation was less, but still significant, at lower temperatures. The underlying cause was formation of an intermolecular disulphide bond by a thiol-disulphide exchange. Ovalbumin, glucose oxidase and beta-lactoglobulin underwent similar aggregation under the same conditions, but by a somewhat different mechanism.[17]

Much light has been shed on the covalent processes leading to enzyme inactivation. Ahern and Klibanov examined the covalent hanges occurring during inactivation of hen egg lysozyme at 100°C in the pH range 4-8.[18] Much of the starting activity could be recovered if cooling to 25°C promptly followed the 100°C heating, showing that reversible conformational processes occur at first. Longer incubations at 100°C resulted in progressively less recovery of activity, indicating that other processes were also at work. Such inactivation followed first order kinetics and was not due to aggregation. Denaturing solutes were used to distinguish between conformational inactivations. Addition of denaturants allows the protein to emerge gradually from incorrect, scrambled structures formed on heating. Kinetic barriers preventing readoption of the correct tertiary structure may be overcome under these conditions. Hence, if denaturant addition increases resistance to irreversible thermoinactivation, conformational inactivation is occurring. If denaturants afford no protection, the inactivation events are covalent.

Three distinct types of covalent alteration were demonstrated in inactivated lysozyme fractions. Deamidation of asparagine residues was significant at pH 4, pH 6 and pH 8. At pH 6 it accounted entirely for the observed inactivation rate. Peptide bond hydrolysis on the C-terminal side of aspartic acid residues accounted for some of the activity loss at pH 4 only. Destruction of cystine residues (i.e. of -S-S- bonds) could be demonstrated at pH 8 but was not a major component of the inactivation process. At pH 4 and pH 6, covalent processes accounted entirely for the observed activity loss. At pH 8, however, formation of incorrect structures (conformational inactivation) was as important as covalent events in the overall inactivation process (see Fig. 2.1).

Similar covalent processes occurred in the 90°C inactivation of beef pancreas RNase. Here, however, the major inactivating event at neutral pH was disulfide interchange, leading to formation of scrambled disulfide bonds. Thiol inhibitors stabilized the enzyme. Destruction of cystine residues (-S-S- bridges) was also notable and loss of lysine residues was also observed. Loss of both types of residue may be due to a beta-elimination reaction. Deamidation of Asn and/ or Gln residues took place at neutral pH but peptide bond hydrolysis did not occur. Both of these latter processes became much more significant at pH 4, however.[19]

These discoveries have implications for the design of thermostable enzymes by protein engineering. Incorporation of extra disulfide bonds may not always be appropriate[19] (see chapter 8). Indeed, the beta-elimination of cystines occurs widely. More than 12 proteins underwent beta-elimination of disulfide bonds at strikingly similar rates at 100°C, indicating that the rate of reaction is independent of protein structure. The beta-elimination reactions also facilitated deleterious disulfide interchanges.[20] A similar approach has been used in inactivation studies of alpha-amylases from two *Bacillus* species, *B. amyloliquefaciens* (mesophilic) and *B. stearothermophilus* (thermophilic).[21,22]

Deamidation of Asparagine Residues

Klibanov's and other groups have shown that deamidation of asparagine (and, to a lesser extent, of glutamine) is a significant covalent reaction in protein inactivation.[18,21,23] The deamidation reaction is an intramolecular, non-enzymic process. In the case of trypsin at least, the protein conformation in the region of the labile asparagines (i.e. the tertiary structure) influences which asparagines are particularly prone to deamidation.[23]

In terms of primary structure, the identity of amino acid on the C-terminal side of the labile asparagine can greatly influence the rate of reaction.[24] Using model peptides, Geiger and Clarke investigated the deleterious deamidation, isomerization and racemization reactions undergone by Asp and Asn residues. The hexapeptide Val-Tyr-Pro-Asn-Gly-Ala, a sequence within adrenocorticotrophic hormone, was a particular focus of attention. The Asn of this peptide deamidates rapidly under mild conditions, with a half-time of only 1.4 days at 37°C, pH 7.4, and giving rise to an aspartyl succinimide ring structure (Fig. 2.1). The succinimide

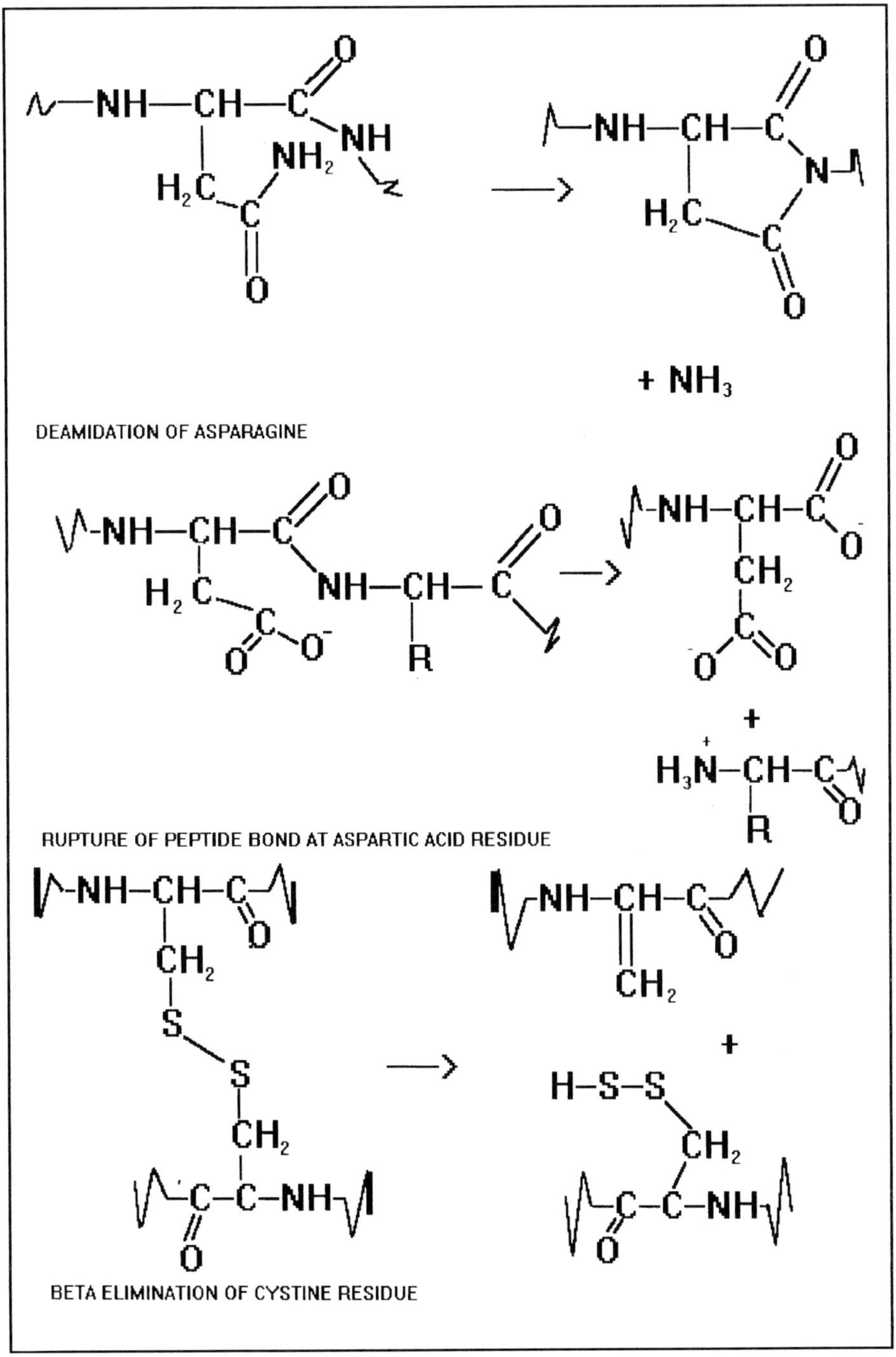

Fig. 2.1. Molecular events leading to loss of protein function.

ring undergoes further reactions: either hydrolysis (half-time 2.3 hours) or racemization (half-time 19.5 hours) may occur. Thus, the deamidation reaction leads to L- and D- (racemized) normal aspartyl and beta-transpeptidated isoaspartyl peptides. (The latter isomerization is where the peptide backbone transfers from the Asp alpha-carboxyl function to its beta-carbon (R-group) carboxyl). Where Asp replaced Asn, the rate of succinimide formation dropped 34-fold. A notable incidence of racemization occurred in tandem with the deamidation and isomerization reactions, and most of this was due to rapid racemization of the succinimide ring. Replacement of the Gly on the C-terminal side of the labile Asn by the much more bulky Leu or Pro decreased the rate of degradation 33 to 50-fold. Indeed, Geiger and Clarke suggest that Asn (and Asp) sites may be 'hotspots' for nonenzymatic degradation over long time periods in vivo.[24] Later work on the peptide Val-Tyr-Pro-X-Y-Ala (where X=Asp, Asn or the alpha- or beta-methyl ester of aspartic acid, and Y=Gly, Ser or Ala) demonstrated that Asn residues formed succinimides 13-36 times more rapidly than Asp residues. Again, the identity of the amino acid residue on the C-terminal side of the Asp/Asn greatly influenced the rate of succinimide formation: the rate decreased in the order Gly > Ser > Ala.[25] Tyler-Cross and Schirch also explored the deamidation of Asn residues in short peptides. They varied the amino acid on *either* side of the central Asn in the pentapeptides Val-Ser-Asn-X-Val and Val-X-Asn-Ser-Val, where X stands for one of ten amino acids. The amino acid residue on the amino side of the Asn had scarcely any effect on the deamidation rate at neutral and alkaline pH. The carboxyl side, however, exerted a significant effect. Glycine, as noted by previous workers, gave the highest deamidation rate (half-time 5.8 hours) but rates were observed to decrease by as much as 70-fold depending on the size and branching of the side chain C-terminal to the Asn. In acidic solution, the deamidation rate was unaffected by the flanking amino acids. No isoAsp residues were detected at acidic pH but the isoAsp:Asp ratio was about 3:1 under neutral and alkaline conditions.[26] Could the labile nature of the asparagine residue be the reason why this is the main site of protein glycosylation? The covalent linkage between the Asn-N and the carbohydrate moiety prevents exposure of an unprotected Asn to a possible deamidation reaction. Glycosylation of proteins can, in itself, have significant stabilizing effects: see the description of the polypeptide hormone erythropoietin (EPO)[27] in later paragraphs of this chapter and see also chapter 6.

Isomerization of Proline

Proline is an imino acid with a rigid 5-membered ring structure as its R-group. A peptide bond preceding a Pro residue (X-Pro, where X is any amino acid) can undergo *cis-trans* isomerization quite readily, unlike peptide bonds preceding other residues. The interconversion has a half-time of 20 minutes at 0°C but its rate increases 3.3-fold for each 10°C rise in temperature.[28] It can, therefore, become a significant event at elevated temperatures, leading to conformational disruption of the polypeptide. It may also prevent correct refolding.[28]

Destructive Oxidation Events

Enzymes and proteins may also undergo oxidation reactions; the cysteine residue is particularly vulnerable and also undergoes destructive thermal reactions.[29] This is a particular hazard for detergent enzymes: commercial formulations, especially in Europe, may contain oxidizing bleaches. The most common one in household detergents is a combination of sodium perborate ($NaBO_3$) with the activator tetra-acetylethylenediamine (TAED). These generate peracetic acid in the wash. Cysteine and methionine (both containing sulfur) are especially prone to destructive oxidation while histidine, tryptophan and tyrosine (all aromatic) react at lesser rates.[30] Misset has described the kinetics of hydrogen peroxide oxidation of the Met216 residue within the detergent protease Maxacal.™ Peroxide oxidation leaves Maxacal with a low-level residual activity. The reaction is second order, since it depends on the concentrations both of enzyme and of peroxide. The peroxide concentration, however, greatly exceeds that of the enzyme (typically by about 1,000-fold). Under these conditions, the reaction becomes pseudo-first order with an offset term for the residual activity of the oxidized enzyme:

$$\underset{(100\%\text{Initial})}{\text{Native Maxacal}} \xrightarrow{k_{ox}[H_2O_2]} \underset{(8-9\%\text{ Offset})}{\text{Maxacal - Met}_{ox}}$$

which leads in turn to the time-dependence equation describing activity at time t (A_t):

$$A_t = \text{Offset} + (\text{Initial} - \text{Offset}).\exp[-k_{ox}.t].$$

Mutation of Met216 to glutamine (or M216Q, where the letters correspond to the single letter amino acid code, the first indicating the wild type residue; the numerals signify position from the N-terminus stabilizes the protease against peroxide oxidation (see chapter 8 for a more detailed description). Diperoxydodecanoic acid (DPDDA), however, is a more potent oxidizing agent which, at 10mM concentration, completely inactivates Maxacal within 20 minutes and reduces the activity of its M216Q variant by 50% within 60 minutes. DPDDA is thought to oxidize a second site besides Met 216 within the protease molecule. DPDDA oxidation of this second site also obeys pseudo first order kinetics but without any offset term since the inactivation reaction goes to completion:

$$\underset{(100\%\text{ Initial, or }A_0)}{\text{Maxacal M216Q + DPDDA}} \xrightarrow{k^{2\text{ nd order}}} \underset{(0\%\text{ Final})}{\text{Maxacal M216Q}_{ox}}$$

This scheme gives rise to the following equation:

$$A_t = A_0 \cdot \exp[-k^{2\text{nd order}} \cdot [\text{DPDDA}] \cdot t].$$

In this case, a replot of the pseudo first order rate constant values against DPDDA concentration allowed estimation of the second order rate constant (0.29 $M^{-1}.sec^{-1}$ at 40°C, pH 9.0).[30]

Proteolytic Processes

The initial, limited proteolytic cleavage(s) of polypeptide molecules occur(s) at exposed and/or flexible loops in the three-dimensional conformation and never within alpha-helices. Fontana's group has shown this time and again for many different proteins, including albumin, calmodulin, citrate synthase, deoxyribonuclease A, diphtheria toxin, elastase, flavocytochrome b_2, hemocyanin, human growth hormone, immunoglobulin, metallothionein, phosphorylase, ribonuclease, serum transferrin, staphylococcal nuclease, subtilisin BPN', thermolysin, thioredoxin, trypsin and tryptophan synthase.[31] Fontana notes that others have associated the initial proteolysis sites with characteristics such as surface accessibility or protrusion but argues that all of these features comprise strongly correlated protein surface characteristics.[31] Proteolysis can occur at exposed/flexible sites in the native polypeptide (N, see Fig. 1.3) or within the unfolded/denatured (U) molecule. There are examples where the T_m index of thermal stability correlates with resistance to proteolysis. At first glance this correlation suggests that it is only the unfolded polypeptide that suffers proteolysis. However, it does not preclude initial 'nicking' of the protein at an exposed or flexible loop without affecting the unfolding equilibrium of the protein. Fontana et al suggest that stabilization of a protein that is prone to proteolytic inactivation could be carried out by introducing mutations to stabilize the folded state. This can be successful for some proteins where T_m values correlate with resistance to proteolysis. In other cases, however, this approach will prove fruitless, but identification of sites of proteolytic attack and their removal by mutation can lead to an improved, resistant polypeptide. Braxton and Wells[32] successfully implemented such a strategy for subtilisin BPN'; their approach is outlined in chapter 8.

Probing the Stabilizing Interactions in Proteins

One can see from the paragraphs above that protein scientists have identified the events responsible for loss of protein function, but what of the molecular interactions *maintaining* a protein's folded structural integrity? (These interactions have been discussed in detail by Dill[33] for proteins in general.) Site directed mutagenesis permits the replacement of specific amino acids within a polypeptide chain. Use of these specific mutagenic techniques allows one to ascertain a particular amino acid's contribution to the overall structural integrity and stability of the protein within which that amino acid is located. The introduction (or elimination) of specific interactions in a protein is now possible. Experiments of this sort have greatly increased understanding of the types of forces contributing to stability.[34,35] Examples of the sorts of substitutions possible include replacement of tyrosine by phenylalanine to assess the contribution of the tyrosine -OH group, or exchange of aspartic with glutamic acid to extend or shorten the position of the carboxylic acid function by a distance of one -CH_2- group. Non-standard amino acids (such as L-3-iodotyrosine,[36] (S)-p-nitrophenylalanine, homophenylalanine or R-3-amino-2-benzylpropionic acid[37]) can also be inserted at desired locations.[36,37] A number of monographs are available which describe the wide scope of site directed mutagenesis and include sections dealing specifically with the issue of stability.[38-40] Examples of purposeful manipulations of a variety of proteins to improve their stability characteristics are outlined in chapter 8. There follows a short survey of the denaturation of specific mutant proteins in comparison with their corresponding wild types.

Matthews has reviewed the effects of various substitutions on the folded stability of T4 lysozyme. Many surface residues can be replaced with alanine with minimal effects on stability. T4 lysozyme has a net positive charge of 9 at neutral pH, suggesting that considerable repulsion of like charges should occur. Substitution of particular positive lysine or arginine residues by negative glutamic acid residues led to very little change (compared with the wild type) either in thermal stability or in crystal structure. Thermal stabilties resembled the wild type's even on construction of double, triple or quadruple mutants. These findings suggest that long range electrostatic interactions contribute little to this protein's stability. Similar results emerged from four attempts to introduce additional stabilizing salt bridges. Again the mutant crystal structures closely resembled the wild type, showing that little if any change had been induced in the protein's folding geometry. The strength of a salt bridge will, of course, depend on pH. At very low pH values the acidic residue of the salt bridge will lose its negative charge and the salt bridge as such will no longer exist. Stability of three of the mutants tested (T115E, N116D and N144E, where the first letter within each mutant designation identifies the wild-type residue by the single-letter amino acid code, the number denotes the position from the N-terminus and the end letter signifies the mutant, substituting amino acid at the indicated position) showed a pH dependence, as would be expected for a salt bridge. However, a more rigorous analysis of mutants, in which the positive partner (lysine or arginine) was replaced by methionine, showed that each mutant retained stability at pH 6.5, even though the positive half of the salt bridge had been removed. Further work demonstrated that the stability increases noted in all three cases were in fact due to favorable interactions with alpha-helix dipoles.[41]

An alpha-helix dipole arises from the similar orientation within an alpha-helix of the constituent polar peptide bonds. The additive effect of these peptide bond dipoles imparts a net dipole to the alpha-helix.[42] This results in a partial positive charge at the N-terminus of the alpha-helix and a partial negative charge at its C-terminus. Such foci of partial charge can lead to undesirable and destabilizing repulsions of like charges. These can be relieved by, for instance, the introduction of a negatively-charged residue (glutamic or aspartic acid) close to an alpha-helix's N-terminus. This consistently increases stability, as proven by at least five examples in T4 lysozyme alone. This sort of stabilization occurs at pH 6.7 but not at pH 2.0, where the carboxyl group is no longer charged. Substitution of the corresponding amide residues (glutamine for glutamic acid or asparagine for aspartic acid) yields no increase in stability. Both of these observations confirm that the effect is electrostatic in nature. Because of this, the exact geometric position of the substitution is not critical, unlike the situation for a hydrogen bond. Such positional flexibility leads to simplified design schemes for stabilizing interactions, providing one takes account of the molecular context in designing the substitution,[41] In general, and especially in light of the above paragraphs, solvent-exposed flexible parts of the protein are less important for stability than are the rigid portions such as alpha-helices or the hydrophobic core.

Other experiments on T4 lysozyme have shown the importance of hydrophobic core packing in protein stability. A protein folds so as to bury as many as possible of its hydrophobic R-groups in the center of the molecule.[43,44] Hydrophobic interactions contribute greatly to the stability of the folded state.[3,45] Leucine is a bulky hydrophobic amino acid. Eriksson and colleagues replaced leucine with alanine in a number of internal locations in T4 lysozyme. The alanine R-group is smaller than that of leucine by $(CH_3)_2$-CH- and creates a cavity within the enzyme core. The individual Leu→Ala mutants were L46A, L99A, L118A, L121A, L133A and also F153A where the phenyl group is replaced by a simple methyl group. Cavity size depends on the exact position of the substitution (L46A yields the smallest cavity, L99A the largest). The Gibbs (free) energy of unfolding decreased with increasing cavity size: stability of the double mutant L99A/F153A decreased by an amount roughly equal to the sum of the individual destabilizations.[41,46] Matthews has explained how it may be possible to separate the experimentally-observed 'cavity destabilization' into constant and variable portions. The 'constant' part is due to the difference in hydrophobicity (expressed as energy of transfer from water to octanol) between the wild type leucine and mutant alanine residues. It approximately equals 1.9 $kcal.mol^{-1}$ and varies with the identities of the wild type amino acid and its replacement. The 'variable' portion depends on the volume of the introduced cavity.[41,46] These findings emphasize the importance for protein stability of tight hydrophobic packing of the polypeptide core.

Mendel et al used site specific mutagenesis with non-natural amino acids to replace the wild type leucine at position 133 in T4 lysozyme. Replacements were tailored into this position to measure three distinct phenomena: the 'cost' of stepwise removal of methyl groups from the hydrophobic core, enlarging a cavity within the T4 lysozyme molecule (accomplished using norvaline, ethylglycine and alanine), the effects of side chain solvation (achieved by changing the hydrophobicity of isosteric amino acid R-groups with norvaline and O-methyl serine substitutions) and the effect on protein stability of packing density and side chain conformation (by burying a greater hydrophobic surface area and changing the configurational entropy of R-groups using leucine, *S,S*-2-amino-4-methylhexanoic acid and *S*-2-amino-3-cyclo-pentylpropanoic acid replacements). Incorporation of the non-natural amino acids took place via suppression of a Leu133 → stop codon in vitro with a chemically-acylated suppressor tRNA. Controls were undertaken to ensure that none of the natural amino acids was incorporated under these conditions. Lysozyme with Leu at position 133 (i.e. wild type amino acid but incorporated using the suppressor system just described) had a T_m value of 43.5°C. Stepwise removal of methyl ($-CH_3$) and methylene ($-CH_2-$) groups successively decreased conformational stability but not in a linear manner. The differences in Gibbs energy values increased with decreasing surface area of the side chain: the leucine → norvaline change (loss of a methyl group) decreased conformational stability by 1.1 kcal.mol^{-1} (T_m 39.5°C) while the norvaline → ethylglycine substitution (loss of one further methyl group, i.e., two methyl groups smaller than leucine) resulted in a 2.2 kcal.mol^{-1} loss of stability (T_m 31°C). Methylserine, which is isosteric with norvaline but contains an -O- atom in place of $-CH_2-$, was 1.7 kcal.mol^{-1} less stable than norvaline (T_m 39.5° and 32.9°C, respectively). Methylserine is more likely to be solvated and less likely to be buried in the protein core than the more hydrophobic norvaline. On the other hand, substitution of Leu 133 by *S,S*-2-amino-4-methylhexanoic acid and *S*-2-amino-3-cyclopentylpropanoic acid, which partially fill the cavity adjacent to Leu 133 in T4 lysozyme's structure with, respectively, one extra methyl group and with two extra methylene groups (part of the cyclopentane ring), stabilized the enzyme by 1.9°C (0.60 kcal.mol^{-1}) and 4.3°C (1.24 kcal.mol^{-1}), respectively.[47]

Quite similar conclusions emerged from studies on barnase, a ribonuclease from *Bacillus amyloliquefaciens* ('*B.a.*RNase'). This is one of the smallest enzymes known and undergoes a reversible denaturation via a two-state mechanism. Nondisruptive deletions were engineered into the three-dimensional enzyme structure by site directed mutagenesis.[48] These are mutations which remove defined existing interactions without introducing any new ones. This approach minimizes any perturbation of the Gibbs energy of unfolding. In effect, it means that the replacement amino acid at a particular position should be similar in size to the wild type residue being substituted. Three variants were investigated: I96V (removal of a $-CH_2-$ group), I96A (removal of 3 $-CH_2-$ groups) and F7L (removal of a benzene ring). Using resistance to urea unfolding as the index of stability, it was shown that these mutants were destabilized with respect to wild type barnase by 1.1, 1.6 and 4.6 kcal/mol, respectively.[48] Other work on barnase has concentrated on the stabilizing effects of electrostatic interactions, such as that between a charged histidine residue and an alpha-helix dipole[49] and by effective N-capping of alpha-helices.[50]

Residue 49 of the alpha subunit of *E. coli* tryptophan synthase lies deep within the protein interior. When each of the 20 amino acids were placed at this position, there was a linear increase in stability of the variants with increasing hydrophobicity of the substituted residue. This relationship broke down only where the volume of the substituted residue exceeded certain dimensions.[51]

Sandberg and Terwilliger have discussed these and other findings and propose a model for the stability effects of apolar substitutions within proteins. This takes into account both the effect of increasing hydrophobicity and of protein core packing densities. It appears that each site within a protein has a characteristic size/packing energy.[43]

Burley and Petsko have highlighted an unexpected stabilizing interaction in oligopeptides and proteins, namely polar interactions of aromatic side chains of amino acids.[52] X-ray crystallography and database analysis have suggested that the interior packing of a protein is partly determined by interactions between the electrons of aromatic R-groups. These interactions may be considered electrostatic in origin and have been dubbed weakly polar: each one is thought to contribute 1-2.5 kcal.mol^{-1} to the enthalpic stability of a protein. Their nature is described in ref. 52.

Replacement of Conserved Residues: *E. Coli* Thioredoxin

Conservation of certain amino acid residues among similar proteins across different species is generally regarded as implying an important structural or functional role for the residues in question. In some cases, however, it has proven possible to replace conserved residues and still recover a functional protein. Gleason has described such a study of three conserved residues in *Escherichia coli* thioredoxin.[53] Thioredoxins are redox proteins: they act as reducing agents and as protein disulfide reductases. They contain many disulfides, have about 100 amino acid residues and occur in many different types of organism. The *E. coli* thioredoxin may become a very useful tool: a thioredoxin gene fusion expression system has been described which prevents formation of inclusion bodies in *E. coli* cytoplasm and which may also aid the purification of recombinant proteins.[54] The active site sequence Trp-Cys$_{32}$-Gly-Pro-Cys$_{35}$ is highly conserved, along with a number of close-lying residues. Three of these amino acids were replaced by site-specific mutagenesis: Asp26, Pro34 and Pro76. These occur in most thioredoxins and lie near the active site disulfide. A mutant containing alanine at position 26 (D26A) was more stable than the wild type but did have altered functional properties despite an insignificant change in redox potential. Wild type thioredoxin's guanidine HCl midpoint for unfolding is 3.4 M at pH 7.0; that of the D26A mutant shifted upwards to 4.3 M, corresponding to a 1.0 kcal.mol^{-1} increase in stability. (The D26A variant has also been reported to be 4.6 kcal.mol^{-1} more stable than the wild type, at pH 8.5.)[55] D26A proved to be a poor reducing agent for ribonucleotide reductase and displayed a 10-fold increase over the wild type's K_m as a substrate for thioredoxin reductase. Alteration of Pro34 to serine had little effect on stability (its denaturation curve did not differ from the wild type's) but this variant's 9 mV decrease in redox potential made it a better reducing agent for ribonucleotide reductase. The P76A mutation was quite destructive. This mutant was destabilized (the guanidine HCl unfolding midpoint dropped to 1.8 M, a 1.4 kcal.mol^{-1} destabilization), the charge distribution in the active site region was altered and, with a higher redox potential than the wild type's, it was not a useful reducing agent for ribonucleotide reductase. It seems that Pro76 helps maintain thioredoxin's redox potential and its active site conformation.

Carbohydrate Side Chains and Protein Stability

Glycosylation is the covalent addition of carbohydrate residues to R-groups of amino acids within a polypeptide backbone. This post-translational modification occurs widely in eukaryotic systems and generally involves the side chains of Asn ('N-linked') or Ser ('O-linked') residues. Besides having a number of purposes in vivo, this carbohydrate labelling of a protein can confer quite a deal of additional stability. The hormone erythropoietin (EPO), which stimulates the production of red blood cell precursors and is an important therapeutic, shows just how significant this extra stabilization can be. EPO is a 34-38 kDa protein containing three N-linked and one O-linked carbohydrate side chains. Carbohydrate accounts for 40-50% of its molecular mass.[27,56,57] The N-linked oligosaccharides are essential for in vivo activity but the O-linked one is not.[56,57] Removal of sialic acid from the N-linked oligosaccharides abolishes 'in vivo' activity (the incorporation of ^{59}Fe radioactivity into red blood cells of starved rats) even when the rest of the sugar moiety remains.[56] The intact N-linked sugar moieties seem to enable the hormone to reach its target sites, since activity in 'in vitro' bioassays (an ELISA protocol and incorporation of tritiated thymidine into mouse fetal liver or bone marrow cells) remains after desialylation and even after complete N-deglycosylation.[56,57]

Tsuda and colleagues incubated intact recombinant human EPO (expressed in baby hamster kidney, or BHK, cells) and EPO fractions treated with various glycosidases at 70°C and measured residual activities using an in vitro method (tritiated thymidine inorporation). All of the glycosidase-treated EPOs were much less stable than the intact molecule and the fully deglycosylated form was least stable of all. Undigested (native) EPO fully tolerated 70°C for 15 minutes while fully deglycosylated EPO retained only 11% activity. EPO lacking sialic acid alone remained 35% active under the same conditions.[56] Thus the carbohydrate moieties and especially their terminal, negatively-charged sialic acids contribute greatly to the EPO protein's thermal stability. A group at Amgen studied the carbohydrate stabilization of EPO conformation in some detail.[27] Using near UV circular dichroism (CD) to observe conformational changes, they subjected different EPOs to heat, pH changes and high concentrations of guanidinium chloride. They used human EPO expressed in both *Escherichia coli* (which cannot glycosylate recombinant proteins) and in Chinese hamster ovary (CHO) cells (where glycosylation proceeds normally). A desialylated form of the CHO-expressed EPO was also studied. Both mammalian-expressed forms remained soluble at pH 7.0 throughout incubation at elevated temperatures up to 75°C and showed a T_m at 56°C. In contrast, the *E.coli* form precipitated completely at 44°C. Even in the presence of 35% glycerol, the 'bacterial' EPO precipitated at 55°C. Note that the sialic acids are not required for this carbohydrate stabilization of EPO, since the asialo form behaved in exactly the same way as the intact form.

In guanidine hydrochloride (GdmCl), both 'mammalian' proteins showed cooperative unfolding with a midpoint at 1.75 M GdmHCl while the '*E. coli*' EPO's midpoint of unfolding was at 1.2 M GdmHCl, indicating an unfolding Gibbs energy 1.4 kcal.mol^{-1} less than that of the glycosylated forms. Unfolding was reversible for both the CHO- and *E.coli*-expressed EPOs. The carbohydrate moieties seem to stabilize EPO against GdmHCl unfolding and, again, the sialic acids are not necessary for this protection. The *E.coli*-derived molecule was less resistant to pH-induced conformational changes than either CHO-expressed EPO. The latter proteins had a transition midpoint at pH 3.75 in the absence of glycerol and at pH 3 in the presence of 35% glycerol. The 'bacterial' EPO showed a transition midpoint at pH 4.4

when 35% glycerol was present. Without it, the *E. coli* material precipitated at or below pH 5.5. Far UV CD spectra indicated that all three proteins retained most of their secondary structure even at pH 2.5. ANS (1-anilino-8-naphthalene sulfonic acid) is a fluorescent molecule which binds to hydrophobic regions of proteins. In binding experiments on 'mammalian' and 'bacterial' EPOs, more ANS bound to the *E. coli* form than to either CHO-expressed form. This implies a greater number of surface hydrophobic sites on the unglycosylated EPO, which are obscured by carbohydrate attachment. The exposed sites of the *E. coli* form are also more hydrophobic than those on the CHO-expressed EPOs: the ANS fluorescent emission peak is shifted to shorter wavelengths. The presence or absence of sialic acid has little effect on ANS binding to the glycosylated forms.[27]

There is an apparent difference between the results of the Tsuda[56] and Amgen groups,[27] reflecting the different indices of stability studied. Tsuda et al looked at the 'in vitro' activities of intact, asialo and deglycosylated EPOs. (Remember that sialic acid must be present for 'in vivo' activity.) In contrast, the Amgen group observed unfolding of the EPO molecule by CD spectroscopy. Thus, Tsuda et al reported losses after 15 minutes incubation at 70°C of 0, 65 and 90% for intact, asialo and deglycosylated EPO respectively. Narhi and her Amgen colleagues note, however, that CD results indicate complete unfolding of EPO by this treatment. They suggest that the observed 'in vitro' activities are probably due to refolding of the molecule. The differences found by Tsuda's group among the three EPO types probably reflect different reversibilites of the heat denaturation. While sialic acid does not affect the stability of the fully folded EPO molecule, it may assist in the reversal of heat-induced unfolding. Indeed, the observed carbohydrate stabilization of EPO could arise from destabilization of the unfolded state by the bulky sugar moieties.[27]

Renin is another medically-relevant example of the importance of glycosylation to protein stability. Renin is a proteolytic enzyme with a role in the angiotensin homeostasis system of the kidney. Renin occurs in mouse submaxillary gland with two different phenotypes, the so-called 'high producers' and 'low producers'. The low producers have a single renin gene ren 1 coding for the glycosylated protein renin-1. In contrast, the high producing strains have two renin genes: Ren 1, which codes for the renin-1, and Ren 2, coding for an unglycosylated and less thermostable renin, renin-2, which lacks three glycosylation sites in its primary sequence. Mutagenesis in vitro, together with renin cleavage of angiotensin I, were used to determine if the observed difference in thermal stability between renins 1 and 2 was actually due to the presence or absence of carbohydrate. Two potential N-glycosylation sites (Asn-X-Ser/Thr) were engineered into the renin-2 protein at the amino acid trios beginning at Asp68 and Asn138, homologous to the first two glycosylation sites of renin-1. The resulting glycosylated renin-2 mutant was much more stable than wild type renin-2 on incubation at 60°C. Endoglycosidase F treatment of the renin-2 mutant cleaves the Asn-sugar links and this treatment restored the mutant's heat stability characteristics to those of the wild type renin-2, showing that the stabilization is due to the carbohydrate moieties and not to the substituted amino acids in the primary sequence. The doubly-glycosylated renin-2 was not as stable as renin-1 and this difference probably reflects the influence of the third sugar moiety of renin-1.[58]

Horseradish peroxidase (HPR) is widely used as an indicator enzyme in analytical biology. It is especially valuable in DNA probes and in immunoassays.[59] It also has uses in organic chemistry[60] and, potentially, in waste treatment.[61] The glycoprotein HRP isoenzyme C has been well characterized[62] and its gene has been cloned and expressed.[63,64] Egorov et al reported Chiswell and Ortlepp's *E. coli*-expressed, unglycosylated, recombinant HRP[63] to be 3 fold less stable at 56°C than commercial plant enzyme.[65] Bonnaffe et al found that periodate-deglycosylated plant HRP was 2.8 times less stable at 50°C than untreated enzyme[66] while Hiner et al noted that carbohydrate-free recombinant HRP C was twice as sensitive as plant HRP to H_2O_2 inactivation.[67]

Is There a Trade-Off Between Stability and Activity?

Studies of denaturation and of stabilizing interactions in proteins strongly suggest that proteins are optimized for function (e.g. an enzyme as catalyst for its specific reaction) rather than for stability (with the obvious exception of the proteins occurring in extremophilic organisms). There may be a trade-off between activity (which requires some degree of molecular flexibility; cf. the induced fit hypothesis) and folding stability (which demands a rigid polypeptide backbone). Indeed, Shoichet et al have stated this view in a most radical form: 'protein residues that contribute to catalysis or ligand binding are not optimal for stability' and they opine that one should be able to substitute functionally-important residues so as to gain stability at the price of reduced activity.[68] Convincing evidence for this view comes from their study of active site substitutions in T4 lysozyme. The catalytic residues Glu11 and Asp20 were each replaced by other residues. Glu11 was substituted by Phe, Met, Ala, His, Asn and Asp20 was substituted by Asn, Thr, Ser and Ala. All of these changes dramatically impaired activity; catalytic efficiencies typically decreased by a factor of 10^4 or greater. Some of the substitutions, however, resulted in increased T_m values and notable gains in thermal stability. The E11F mutant, for instance, had a T_m (determined by circular dichroism) 4.3°C greater than the wild type, accompanied by a stabilizing Gibbs energy change of 1.7 kcal.mol^{-1}. Similarly, the D20N variant yielded gains of 3.1°C and 1.3 kcal.mol^{-1} in T_m and Gibbs energy, respectively. Changes to two of the residues involved in substrate binding (Ser117 and Asn132) were also stabilizing: S117V was 5.1°C and 2.0 kcal.mol^{-1} more stable than the wild type. This particular mutant had 5% of wild type activity, so activity losses occurred at these positions also. Substitutions of Asn132 combined stability gains of 1.2-1.5 kcal.mol^{-1} with activity reductions of only 2.5- to 5-fold. (Note, however, that some of the mutants prepared were actually *de*stabilized.) Crystal structures of the stabilized variants D20N, D20S and of the destabilized mutants D20A, E11N, E11H showed little net change in the T4 lysozyme structure. However, more extensive structural changes occurred in the stabilized E11M and E11A forms. Both of these replacements displayed a 'hinge-bending' motion of the protein: the active site had been partially closed by the two domains moving towards each other. Both Met and Ala are non-polar residues; mutants with polar residues at position 11 (E11N, E11H) did not show this active site closure. A decreased polar surface area in the stabilized mutants may explain the 'hinge-bending' motion.[68] Interestingly, some of the active site mutations were more stabilizing than previously-discovered point mutations.[68,41] It seems that the stabilizing alterations improve the complementarity of R-group interactions within the lysozyme polypeptide but, at the same time, decrease the substrate/transition state complementarity (and hence catalytic activity). Shoichet and colleagues cite a number of other studies in the literature which point towards the idea of stability/activity compromise. They also suggest that the need to reconcile these two requirements may explain the large size and low stability of protein molecules.[68]

In as stark a manner as Shoichet et al, Drexler has stated that 'natural proteins are needlessly destabilized'.[69] He argues that unfolding Gibbs energy values in the range 5-20 kcal.mol^{-1} (ref. 33) mean that the probability of a protein's being in the unfolded state is 10^{-4}-10^{-15}. Since the time required for protein folding ranges 10^{-1}-10^{3} seconds,[28] one might have to wait longer than 100 years to observe the unfolding of an intact protein with a stability of 15 kcal.mol^{-1}. Few creatures survive this long, so evolution will not operate to favor more stable proteins over 'ordinary' ones. It is likely, of course, that proteins may evolve to withstand some in vivo denaturing/inactivating influence or event[69] and natural selection must surely have favored the adaptation of extremophiles to their current habitats.

Conclusion

From the above examples, one can state that certain conclusions have emerged from studies of native and mutant proteins under extreme conditions. Inactivating covalent processes, which follow unfolding of the polypeptide, fall into a small number of defined reactions. Electrostatic interactions can be stabilizing in appropriate, rigid regions of the polypeptide. However, tight packing of the hydrophobic core so as to minimize cavities is perhaps the most important contributor to folded protein stability.

The observations described in this chapter can provide crude initial ground rules for protein stabilization (cf. chapter 8) but they are based on studies in *aqueous* systems. Chapter 3 discusses insight into protein stability derived from study of enzymes and other proteins in *organic* systems.

References

1. Sadana A. Enzyme deactivation. Biotechnol Adv 1988; 6:349-466
2. Mateo PL and Privalov PL. Pepsinogen denaturation is not a two-state transition. FEBS Lett 1981; 123:189-192.
3. Daniel RM, Dines M and Petach HH. The denaturation and degradation of stable enzymes at high temperatures. Biochem J 1996; 317:1-11.
4. Becktel WJ and Schellman JA. Protein stability curves. Biopolymers 1987; 26: 1859-1877.
5. Privalov PL, Griko YuV, Venyaminov SYu et al. Cold denaturation of myoglobin. J Mol Biol 1986; 190:487-498.
6. Dill KA, Alonso DOV and Hutchinson K. Thermal stabilities of globular proteins. Biochemistry 1989; 28:5439-5449.
7. Privalov PL. Cold denaturation of proteins. CRC Crit Rev Biochem Mol Biol 1990; 25:281-305.
8. Mozhaev VV and Martinek K. Inactivation and reactivation of proteins (enzymes). Enzyme Microb Technol 1982; 4:299-309.
9. Johnson RE, Adams P and Rupley JA. Thermodynamics of protein cross-links. Biochemistry 1978; 17:1479-1484.
10. Pace CN, Shirley BA and Thomson JA. Measuring the conformational stability of a protein. In: Creighton TE, ed. Protein Structure: a Practical Approach. Oxford: IRL Press, 1989:311-330.
11. Zale SE and Klibanov AM. On the role of reversible denaturation (unfolding) in the irreversible thermal inactivation of enzymes. Biotechnol Bioeng 1983; 25: 2221-2230.
12. Hoare M, Kahn MR and Dunnill P. Interfacial damage to proteins during intensive mixing in fermentation and downstream processing. In: Van den Tweel WJJ,

Harder A and Buitelaar RM, eds. Stability and Stabilization of Enzymes. Amsterdam: Elsevier, 1993:21-27.
13. Lee YK and Choo CL. The kinetics and mechanism of shear inactivation of lipase from *Candida cylindracea*. Biotechnol Bioeng 1989; 33:183-190.
14. Bowen WR and Gan Q. Properties of microfiltration membranes: the effects of adsorption and shear on the recovery of an enzyme. Biotechnol Bioeng 1992; 40:491-497.
15. Mozhaev VV. Mechanism-based strategies for protein thermostabilization. Trends Biotechnol 1993; 11:88-95.
16. Quax WJ. Thermostable glucose isomerases. Trends Food Sci Technol 1993; 4:31-34.
17. Liu WR, Langer R and Klibanov AM. Moisture-induced aggregation of lyophilized proteins in the solid state. Biotechnol Bioeng 1991; 37:177-184.
18. Ahern TJ and Klibanov AM. The mechanism of irreversible enzyme inactivation at 100°C. Science 1985; 228:1280-1284.
19. Zale SE and Klibanov AM. Why does ribonuclease irreversible inactivate at high temperatures? Biochemistry 1986; 25: 5432-5444.
20. Volkin DB and Klibanov AM. Thermal destruction processes in proteins involving cystine residues. J Biol Chem 1987; 262:2945-2950.
21. Tomazic SJ and Klibanov AM. Mechanisms of irrreversible thermal inactivation of *Bacillus* alpha-amylases. J Biol Chem 1988; 263:3086-3091.
22. Tomazic SJ and Klibanov AM. Why is one *Bacillus* alpha-amylase more resistant against irreversible thermoinactivation than another? J Biol Chem 1988; 263: 3092-3096.
23. Kossiakof AA. Tertiary structure is a principal determinant to protein deamidation. Science 1988; 240:191-194.
24. Geiger T and Clarke S. Deamidation, isomerization and racemization at asparaginyl and aspartyl residues in proteins. J Biol Chem 1987; 262:785-794.
25. Stephenson RC and Clarke S. Succinimide formation from aspartyl and asparaginyl peptides as a model for the spontaneous degradation of proteins. J Biol Chem 1989; 264:6164-6170.
26. Tyler-Cross R and Schirch V. Effects of amino acid sequence, buffers and ionic strength on the rate and mechanism of deamidation of asparagine residues in small peptides. J Biol Chem 1991; 266: 22549-22556.
27. Narhi, LO, Arakawa T, Aoki K et al. The effect of carbohydrate on the structure and stability of erythropoietin. J Biol Chem 1991; 266:23022-23026.
28. Creighton TE. Proteins: Structures and Molecular Properties. NY: WH Freeman, 1990:181, 311-312.
29. Volkin DB and Middaugh CR. The effect of temperature on protein structure. In: Ahern TJ and Manning MC, eds. Stability of Protein Pharmaceuticals, Part A. Chemical and Physical Pathways of Protein Degradation. NY: Plenum Press Inc., 1992:215-247.
30. Misset O. Stability of industrial enzymes. In: Van den Tweel WJJ, Harder A and Buitelaar RM, eds. Stability and Stabilization of Enzymes. Amsterdam: Elsevier, 1993:111-131.
31. Fontana A, Polverino de Laureto P and De Filippis V. Molecular aspects of proteolysis of globular proteins. In: Van den Tweel WJJ, Harder A and Buitelaar RM, eds. Stability and Stabilization of Enzymes. Amsterdam: Elsevier, 1993: 101-110.
32. Braxton S and Wells JA. Incorporation of a stabilizing Ca^{2+}-binding loop into subtilisin BPN'. Biochemistry 1992; 31: 7796-7801.
33. Dill KA. Dominant forces in protein folding. Biochemistry 1990; 29: 7133-7155.
34. Alber T. Mutational effects on protein stability. Ann Rev Biochem 1989; 58: 765-798.
35. Beasty AM, Hurle M, Manz JT et al. Mutagenesis as a probe of protein folding

and stability. In: Oxender DL, Fox CF, eds. Protein Engineering. NY: Alan R Liss, 1987:91-102.

36. Benner SA. Expanding the genetic lexicon: incorporating non-standard amino acids into proteins by ribosome-based synthesis. Trends Biotechnol 1994; 12: 158-163.
37. Anthony-Cahill SJ, Griffith MC, Noren CJ et al. Site-specific mutagenesis with unnatural amino acids. Trends Biochem Sci 1989; 14:400-403.
38. Oxender DL and Fox CF, eds. Protein Engineering. NY: Alan R Liss, 1987.
39. Nosoh Y and Sekiguchi T. Protein Stability and Stabilization Through Protein Engineering. Chichester UK: Ellis Horwood, 1991.
40. Rees AR, Sternberg SJE and Wetzel R. Protein Engineering: a Practical Approach. Oxford:IRL Press, 1992.
41. Matthews BW. Structural and genetic analysis of protein stability. Annu Rev Biochem 1993; 62:139-160.
42. Hol WGJ, van Duijnen PT and Berendsen HJC. Alpha-helix dipole and properties of proteins. Nature 1978; 273:443-446.
43. Sandberg WS and Terwilliger TC. Repacking protein interiors. Trends Biotechnol 1991; 9:59-63.
44. Mozhaev VV and Martinek K. Structure-stability relationships in proteins: new approaches to stabilizing enzymes. Enzyme Microb Technol 1984; 6:50-59.
45. Baldwin EP and Matthews BW. Core-packing constraints, hydrophobicity and protein design. Curr Opin Biotechnol 1994; 5:396-402.
46. Eriksson AE, Baase WA, Zhang X-J et al. Response of a protein structure to cavity-creating mutations and and its relation to the hydrophobic effect. Science 1992; 255:178-183.
47. Mendel D, Ellman JA, Chang Z et al. Probing protein stability with unnatural amino acids. Science 1992; 256: 1798-1802.
48. Kellis JT, Nyberg K, Sali D et al. Contribution of hydrophobic interactions to protein stability. Nature 1988; 333:784-786.
49. Sali D, Bycroft M and Fersht AR. Stabilization of protein structure by interaction of alpha-helix dipole with a charged side chain. Nature 1988;335:740-743.
50. Serrano L and Fersht AR. Capping and alpha-helix stability. Nature 1989; 342:296-299.
51. Yutani K, Ogasashara K, Tsujita T et al. Dependence of conformational stability on hydrophobicity of the amino acid residue in a series of variant proteins substituted at a unique position of tryptophan synthase alpha subunit. Proc Nat Acad Sci USA 1987; 84:4441-4444.
52. Burley SK and Petsko GA. Electrostatic interactions in aromatic oligopeptides contribute to protein stability. Trends Biotechnol 1989; 7:354-359.
53. Gleason FK. Mutation of conserved residues in *Escherichia coli* thioredoxin: effects on stability and function. Protein Sci 1992; 1:609-616.
54. La Vallie ER, Di Blasio EA, Kovacic S et al. A thioredoxin gene fusion expression system that circumvents inclusion body formation in the *E. coli* cytoplasm. Bio/ Technology 1993; 11:187-193.
55. Langsetmo K, Fuchs JA and Woodward C. The conserved, buried aspartic acid in oxidized *Escherichia coli* thioredoxin has a pK_a of 7.5. Its titration produces a related shift in global stability. Biochemistry 1991; 30:7603-7609.
56. Tsuda E, Kawanishi G, Ueda M et al. The role of carbohydrate in recombinant human erythropoietin. Eur J Biochem 1990; 188:405-411.
57. Higuchi M, Oh-eda M, Kuboniwa H et al. Role of sugar chains in the expression of the biological activity of human erythropoietin. J Biol Chem 1992; 267: 7703-7709.

58. Ladenheim R, Rougeon F. Modification de la thermostabilité de la rénine par sa glycosylation. Comptes Rendu Acad Sci Paris 1993; 316:334-336.
59. Ryan O, Smyth MR and Ó Fágáin C. Horseradish peroxidase: the analyst's friend. In: Tipton KF, ed. Essays in Biochemistry 28. London: Portland Press 1994:129-146.
60. Dordick JS. Enzymatic and chemoenzymatic approaches to polymer synthesis. Trends Biotech 1992; 10:287-293.
61. Miland E, Smyth MR and Ó Fágáin C. Phenol removal by modified peroxidases. J Chem Technol Biotechnol 1996; 67: 227-236.
62. Dunford HB. Horseradish peroxidase: structure and kinetic properties. In: Everse J, Everse KE and Grisham MB, eds. Peroxidases in Chemistry and Biology, vol 2. Boca Raton: CRC Press, 1991:1-24.
63. Chiswell DJ and Ortlepp SA. New synthetic DNA encoding horseradish peroxidase C isoenzyme. European Patent EP 299682: 1989.
64. Smith AT, Santama S, Dacey S et al. Expression of a synthetic gene for horseradish peroxidase C in *Escherichia coli* and folding and activation of the recombinant enzyme with Ca^{2+} and heme. J Biol Chem 1990; 265:13335-13343.
65. Egorov AM, Gazaryan IG, Kim BB et al. Horseradish peroxidase isoenzyme C. A comparative study of native and recombinant enzyme produced by *E. coli* transformants. Ann NY Acad Sci 1994; 721: 73-81.
66. Bonnaffe D, Therisod M, Seris JL et al. Chemical deglycosylation of horseradish peroxidase and surglycosylation using a new glycosylating reagent: effects on catalytic activity and stability. In: Van den Tweel WJJ, Harder A and Buitelaar RM, eds. Stability and Stabilization of Enzymes. Amsterdam: Elsevier, 1993: 223-227.
67. Hiner ANP, Hernández-Ruíz J, García-Cánovas F et al. A comparative study of the inactivation of wild-type, recombinant and two mutant horseradish peroxidase isoenzymes by hydrogen peroxide and and m-chloroperoxybenzoic acid. Eur J Biochem 1995; 234:506-512.
68. Shoichet BK, Baase WA, Kuroki R et al. A relationship between protein stability and protein function. Proc Nat Acad Sci USA 1995; 92:452-456.
69. Drexler KE. Molecular nanomachines: physical principles and implementation strategies. Annu Rev Biophys Biomol Struct 1994; 23:377-405.

CHAPTER 3

Enzymes in Organic Media

Introduction

The successful use of non-membrane bound enzymes in biphasic aqueous-organic systems (and even in anhydrous organic solvents) was a novel and surprising development. Catalysis of unusual reactions can occur in these systems and many such processes have been documented, examples are outlined in refs. 1-6. This is a rapidly-developing field, but a number of good reviews may be found in refs. 7-12. Here the focus will be confined to the stability aspects of such studies.

Enzyme Behavior in Anhydrous Organic Solvents

Some enzymes (and other proteins) have been observed to function perfectly well in non-aqueous media. Indeed, they show some remarkable properties in organic systems: see Klibanov[12] for a brief but very informative review. A protein can retain its correct conformation on transfer from a hydrophilic aqueous to a hydrophobic organic solvent system (provided this is done carefully: see ref. 12). This is because a layer of bound or 'essential' water remains associated with the folded polypeptide, even in a 'completely dry' or lyophilized preparation.[12,13] This minimal layer of water is sufficient to solvate the folded polypeptide. An unfolded polypeptide requires a greater number of water molecules for effective solvation; thus the folding equilibrium is shifted towards the folded form in low-water systems.[8] This shift in equilibrium means that denaturation (unfolding) becomes a much less likely event in low-water hydrophobic systems. This is not the only characteristic of these media benefitting stability. A common feature of the inactivating covalent reactions discussed in chapter 2 is the participation of water.[8,14] Water is present at the extremely high concentration of 55M in aqueous systems. Despite the retention of 'essential' water surrounding a polypeptide in a hydrophobic medium, water's effective concentration is profoundly reduced and water-mediated deleterious covalent reactions are much less likely to occur. Thus, both the unfolding and inactivation events of the conventional $N \leftrightarrow U \leftrightarrow I$ model can be markedly decreased in hydrophobic media.

Klibanov has proposed certain rules to ensure that enzymes will remain active in organic solvents:

1. The solvent should be hydrophobic and show very little affinity for water.
2. Enzymes for use in organic solvents should be lyophilized from solutions with pH values corresponding to the enzyme's optimum pH.
3. The enzyme should be agitated vigorously or sonicated to ensure homogeneous dispersion in the organic solvent.[8]

Stabilizing Protein Function, by Ciarán Ó Fágáin.
© Springer - Verlag and Landes Bioscience 1997.

Why are these rules important? In biphasic systems, emulsions and reversed micelles, the enzyme molecule remains in an aqueous milieu, even though the bulk phase is organic.[8] Enough solvent water remains to hydrate the enzyme molecule and to maintain its correct, functional conformation. Hydrophobic organic solvents are superior to hydrophilic ones in this regard since the latter tend to strip away the aqueous layer or 'shell' surrounding the protein molecules. Hydrophobic residues tend to be buried in the protein core while hydrophilic residues tend to be found on the surface. Replacement of the aqueous shell with water-miscible organic solvent critically disturbs the folding of the protein, leading to reduction of or loss of activity.[8]

An enzyme placed in an organic solvent retains the ionic state it had in the previous aqueous system[8] throughout lyophilization and redispersal in organic phase. Obviously the enzyme will function best if it retains the ionic balance corresponding to its pH optimum. The third rule relates to the fact that, in many cases, proteins are suspended and not truly dissolved in organic solvents. It is important to ensure as finely dispersed a suspension as possible to prevent diffusional limitations on performance. True solution does occur sometimes, however. Mozhaev et al were able to dissolve micromolar amounts of the 'normal', hydrophilic alpha-chymotrypsin in non-polar solvents (*n*-octane, cyclohexane and toluene). The enzyme was fully active in the homogeneous solution obtained (light scattering experiments indicated no particles bigger than 100Å), but was distinguished by a notably high thermostability. Full activity was retained during refluxing in 'dry' solvent at temperatures as high as 126°C, even for several hours. The presence of a small percentage (v/v) of polar solvent, or even <1% (v/v) water, greatly shortened the duration of activity. Three distinct methods were used to dissolve enzyme in solvent. Shaking chymotrypsin powder in octane held at room temperature dissolved the enzyme up to 0.3 micromolar concentration. (Any solid particles were removed by centrifugation or by filtration through a 0.22 micrometer filter.) Solubility increased with temperature and with the enzyme-to-octane ratio. It also depended heavily on the humidity of both the solvent and the enzyme powder. A second, much faster method (approx. 30 minutes) involved the addition of acetone or diethyl ether (both highly volatile) to a suspension of the enzyme in octane. This was followed by evaporation under vacuum at room temperature and led to the dissolution of greater amounts of chymotrypsin. A third method used two connected, refluxing, vessels: one contained an enzyme suspension while the other contained the solvent refluxing under sodium. The vapors were in contact and the chymotrypsin solution was obtained by filtration. With this method, up to 1 micromolar chymotrypsin concentration could be obtained in toluene or cyclohexane. Other proteins could be dissolved in similar manner.[15]

Barone and colleagues measured the thermodynamic stability parameters of ribonuclease A in media of differing water content using differential scanning calorimetry (DSC).[16] In general, T_m increased (accompanied by a 2-4-fold decrease in enthalpy values) with decreasing water content in a non-polar solvent such as dodecane. More complex behavior was observed in polar solvents (dimethylformamide or N-methylacetamide; see below). Different effects were noted depending on the solvent and the quantity of water in it. Enthalpy studies on model dipeptides in different solvents suggested that water actually is a weak destabilizer of protein structure. Battistel and Bianchi also studied ribonuclease A in water-immiscible organic media by DSC, but in this case the protein was immobilized on Celite or Amberlite XAD-7 (slightly hydrophobic) to prevent intermolecular interactions and/or aggregation.[17] 'Dry' ribonuclease A on Celite (i.e. adsorbed in the absence of any solvent) underwent a single cooperative unfolding transition (T_m 68°C), the characteristics of which altered little on addition of aqueous solvent. Adsorption to Amberlite, in contrast, decreased the T_m to 51°C, even in aqueous buffer. Also, two transitions were noted, indicating that the unfolding on Amberlite was not two-state (see chapter 1). Celite was used for further experiments owing to its more 'orthodox' behavior. The type of solvent and the degree of protein hydration strongly influenced the unfolding process. Hydrophobic solvents (dodecane and especially octanol) increased the T_m. Using these and other solvents, attempts were made to correlate ribonuclease A stability with various bulk solvent parameters (log P, Hildebrand solubility parameter, H-bonding parameter, dipole moment and dielectric constant). Enthalpy changes during unfolding in different solvents (at a constant 17.5% water content) indicated that the best correlation with protein thermostability was that for solvent dielectric constant. This implies that electrostatic forces contribute significantly to protein stabilization in organic solvents.

Some Case Studies

Some anhydrous-phase stabilizations of important hydrolases and other enzymes will now be considered.

Lipase has been much studied in hydrophobic systems, since its substratres are themselves hydrophobic. Pig pancreatic lipase was studied in a tributyrin system where it catalyzed transesterifications between tributyrin and primary and secondary alcohols.[18] Even at a water concentration of 0.015% the enzyme was catalytically active. Lipase as 'dry' was shown to absorb some of the available water from the water-equilibrated solvent while a 'wet' lipase preparation containing 3.6% water lost water to a solvent mixture containing 0.015% water, suggesting that water can equilibrate between enzyme and organic medium. The 'organic' lipase preparations were much more thermostable at 100°C than an aqueous preparation, which completely lost activity almost at once. Lipase in 0.8% water was 50% active after 15 minutes at 100°C while lipase in 0.015% water had a half life longer than 12 hours at 100°C. The critical water concentration for a 12 hour half life at 100°C seemed to be 0.3%: above this concentration, the 100°C half life decreased markedly. Note that these findings applied to the tributyrin system only; different 100°C half lives occurred in different organic solvent systems.

In the case above, the lyophilized lipase was suspended directly in organic solvent. It was not suspended within reverse micelles or made 'organic soluble' by chemical modification with polyethylene glycol (PEG).[7] Hirata and colleagues investigated transesterifications by *Pseudomonas* lipases in a heterogeneous system of tributyrin and various alcohols. Some of the lipases were very stable in a water-restricted environment: in one case, 25% of initial lipase activity remained after 48 hours' incubation in tributyrin at 90°C. Similar experiments in a toluene system showed that the essential remaining water surrounding the enzyme is stripped away at high temperatures (e.g. 100°C) and that one can restore the activity so lost by supplying water.[19] The increased stability of the 'organic' lipase preparations is ascribed to their decreased conformational mobility in the absence of bulk water. Gross molecular unfolding is greatly diminished and so inactivation rates remain low even at high temperatures.

Subtilisin and alpha-chymotrypsin have likewise been used successfully in organic solvents.[20] The role of these enzymes is to hydrolyze hydrophilic substrates, namely peptides and proteins. The fact that these proteases also can be used in hydrophobic media is significant. It shows that it was not the nature of the lipase enzymes themselves, working as they do on hydrophobic substrates, that permitted them to function in organic solvents. The stability of chymotrypsin at 100°C in dry octane was investigated.[20] The chymotrypsin used had been previously lyophilized from an aqueous solution of pH 7.8 containing the ligand N-acetyl-L-phenylalanine. The half life at 100°C was 4.5 hours. In contrast, the half life of chymotrypsin in aqueous solution is less than 15 minutes even at 55°C. Both sets of measurements took place under conditions that precluded autolysis.

Russell and Klibanov have shown that 'ligand memory' can also influence enzyme conformation in organic media. Subtilisins freeze-dried from aqueous solutions containing specific ligands were much less stable at 110°C in octane than those prepared without ligand. It is thought that the ligands induce a catalytically more active, but less thermostable, conformation of the enzyme. Ligand-influenced subtilisin was also much more sensitive than 'uninfluenced' subtilisin to the denaturing effects of dimethyl sulfoxide.[21]

Zaks and Klibanov propose a kinetic explanation for the greater conformational rigidity of enzymes in organic media.[20] High kinetic barriers prevent the enzyme from adopting an alternative conformation on transfer from an aqueous to an organic solvent. Both the enhanced thermostability and the 'pH memory' of organic-suspended enzymes point to this interpretation. However, thermal inactivation does occur in organic media, albeit at much lower rates than in water-based systems. Studies on three hydrolases (RNase, chymotrypsin and lysozyme) indicated that aggregation was the major cause of inactivation in the systems studied.[22]

Water-Miscible Organic Solvents

The discussion above concerns hydrophobic organic systems. Different considerations apply to water-miscible organic solvents. Because they are freely miscible with water, these solvents tend to strip away the 'essential' water bound to the polypeptide globule. Gorman and Dordick have demonstrated the removal of tritiated water (T_2O) from enzyme surfaces in water-organic cosolvent mixtures.[23] Alpha-chymotrypsin undergoes abrupt threshold changes in catalytic activity and in conformation once the organic solvent concentration increases above a certain critical value. Up to 65% by volume of some organic compounds could be tolerated without any effect on catalytic activity. Above the critical cosolvent concentration, the activity lost could be completely recovered by dilution in aqueous solvent. This shows that the effect is due to reversible denaturation and not to irreversible inactivation (see chapter 2). An entirely different enzyme, the lignin-degrading oxidoreductase known as laccase (EC 1.10.3.2) displayed a threshold dependence of catalytic activity on cosolvent concentration, showing that this effect is not confined either to alpha-chymotrypsin or to hydrolases.[23,24] The denaturing efficiency of organic cosolvents correlated with cosolvent hydrophobicity: a linear log-log plot resulted when the cosolvent concentration yielding 50% of optimal catalytic activity was plotted against the cosolvent partition coefficient (in an octanol/water system). This could have implications for the selection of suitable organic cosolvents for enzymatic reactions.[24] Reslow and colleagues reported similar findings for alpha-chymotrypsin adsorbed on porous glass beads.[25] Further investigations of enzymes in water-cosolvent mixtures noted threshold effects for the reversible denaturation of a number of proteins by a variety of solvents of different nature and have led to the definition of 'denaturation capacity' (DC) as a ranking criterion for solvents to be used in biocatalytic systems.[26] DC is based on a thermodynamic model of reversible solvent-induced protein denaturation and leads to an equation relating an organic solvent's threshold concentration to its physicochemical attributes (e.g. hydrophobicity, solvating ability and molecular geometry). Solvents can be arranged in a scale according to their DC values and one can theoretically predict an organic solvent's threshold concentration for a given protein.[26]

Gupta[27] notes that enzymes have been estimated (e.g. ref. 20) to become active with less than 50 water molecules available for each protein molecule. (See also ref. 13.) He believes that numbers of this order are not accidental. Observable activity in organic solvents may depend only on the complete hydration of charged groups lying on the protein surface. Such a tiny amount of water is enough to break up interactions of charged groups with each other and so 'unlock' an inactive protein conformation. Although it becomes flexible enough to express some activity at this extremely low level of hydration, the enzyme molecule remains very rigid unless and until further amounts of water lead to greater flexibility.[27] Gupta suggests that this rigidity underlies 'pH memory' and some of the other properties of enzymes in anhydrous solvents.[27] However, substances other than water, such as glycerol or formamide, can act to make protein molecules flexible under these conditions. These compounds are capable of forming hydrogen bonds and can replace water in coating the protein surface. Addition of 3% formamide to tyrosinase suspended in octanol led to a notable increase in that enzyme's activity.[13,28]

Other examples of stabilized enzymes in organic media include (+)-alpha-pinene cyclase from sage (*Salvia officinalis*) in hexane[29] and mushroom polyphenol oxidase in chloroform.[30]

Combined Strategies

One can stabilize enzymes for use in the presence of organic solvents by strategies such as immobilization (chapter 7) or protein engineering (chapter 8). Immobilization is reviewed briefly in ref. 10 in the context of organic phase enzyme reactions. The immobilization-protection of alpha chymotrypsin against aqueous/organic solvent mixtures has been described.[31] Alpha-chymotrypsin has also been stabilized for organic solvent use by encagement within crosslinked reverse micelles.[10] Arnold has dealt specifically with the question of protein engineering for organic solvent use, giving due consideration to stability.[11] A subtilisin BPN' site-directed mutant enzyme containing 6 amino acid replacements has been reported to be significantly more stable than the wild type in anhydrous dimethylformamide, but the mutant's catalytic activity was extremely low in this system.[32] An improved variant, dubbed 8397, had a half life of 1600 hours in aqueous solution at pH 8.4 and 25°C compared with 15 hours for wild type subtilisin. In dimethylformamide, wild type enzyme had a half life of 20 minutes while that of 8397 was 350 hours[33] (see chapter 8 for further details).

Membrane-bound (as distinct from soluble) multi-subunit enzymes may also be stabilized as protein-lipid complexes by organic solvents. Ayala and coworkers found that cytochrome oxidase and H^+-ATPase from inner mitochondrial membrane were stabilized by factors of 100 and 9, respectively when the water content of the toluene bulk phase was reduced from 13 microliters per milliliter to 3 microliters per milliliter.[34] (Stability was measured in terms of half life at 70°C.)

The judicious use of hydrophobic, non-polar solvents as alternatives to water has tremendous potential for the achievement of stabilized enzymes. Dramatic increases in thermal and operational stabilities have been achieved by a strategy involving relatively simple procedures (lyophilization and resuspension) which can be carried out quite readily. The enhanced stability occurs without deliberate change of the protein molecule. The increased thermal stability often shown by proteins in organic or cosolvent media is a useful bonus but is also possible to influence the catalytic properties of enzymes in non-aqueous systems. Khmelnitsky et al observed that salts could greatly increase the activities of enzymes suspended in organic solvents. Inclusion of >90% (w/w) KCl (or other salts) in the lyophilized enzyme powder led to notable activation of subtilisin Carlsberg (up to 3750 times) and alpha-chymotrypsin (50-fold) in a transesterification reaction in organic solvent. The activation seemed to arise from an increase in k_{cat} and not from any notable effect on K_m. The activation effect is non-specific with respect to both salt and solvent; it appears to be a feature of the protein-salt matrix. Khmelnitsky and colleagues propose that the activation arises from salt-matrix protection of the enzyme against inactivation from direct contact with organic solvent.[35] Organic solvents facilitate novel enzyme-catalyzed reactions: Khmelnitsky et al measured rates of proteinase-catalyzed *transesterification* in ref. 35, not the 'natural' amidase cleavage of peptide bonds which predominates in aqueous media. Subtilisin has even been used to catalyze reactions in sugar chemistry (the acylation of carbohydrates in anhydrous dimethylformamide).[36] Proteases can also be used to prepare protein conjugates[37] and to accomplish peptide bond synthesis; this last is often favored in organic systems.[38] The ability to use proteolytic enzymes 'in reverse' for peptide bond synthesis in hydrophilic organic solvents is also well documented.[39,40] Lipases can facilitate esterification and transesterification reactions.[41] See refs. 42 and 43 for other examples.

If only because of the relative simplicity and speed factors, organic solvent use must be considered a serious challenger to more involved stabilization strategies. Indeed, as Gupta has pointed out[27] the thermostabilizations achieved in organic media can equal, and even surpass, those resulting from other strategies such as crosslinking, immobilization or even protein engineering. However, current understanding of protein properties in non-aqueous systems allows redesign of a target protein to improve its performance in these media. Such redesign can take place via a strategy of protein engineering (see ref. 11 and chapter 8) or poly(ethylene glycol) coupling.[7]

References

1. Klibanov AM, Cambou B. Enzymatic production of optically active compounds in biphasic aqueous-organic systems. Meths Enzymol 1987; 136:117-138.
2. Carrea G, Cremonesi P. Enzyme-catalyzed steroid transformations in water-organic solvent two-phase systems. Meths Enzymol 1987; 136:150-157.
3. Luisi PL, Steinmann-Hofmann B. Activity and conformation of enzymes in reverse micellar solutions. Meths Enzymol 1987; 136:188-216.
4. Laane C, Hillhorst R, Veeger C. Design of reversed micellar media for the enzymatic synthesis of apolar compounds. Meths Enzymol 1987; 136:216-230.
5. Butler G. Synthesis of urea with urease in water-organic solvent mixtures. Meths Enzymol 1987; 136:234-239.
6. Bender ML. Kinetic studies of immobilized alpha-chymotrypsin in aprotic solvents. Meths Enzymol 1987; 135:537-546.

7. Inada Y, Takahashi K, Yoshimoto T et al. Application of polyethylene glycol-modified enzymes in biotechnological processes: organic solvent-soluble enzymes. Trends Biotechnol 1986; 4:190-194.
8. Klibanov AM. Enzymatic catalysis in anhydrous organic solvents. Trends Biochem Sci 1989; 14:141-144.
9. Dordick JS. Enzymatic catalysis in monophasic organic solvents. Enzyme Microb Technol 1989; 11:194-211.
10. Khmelnitsky YuL, Levashov LV, Klyachko NL et al. Engineering biocatalytic systems in organic media with low water content. Enzyme Microb Technol 1988; 10:710-724.
11. Arnold FH. Engineering enzymes for non-aqueous solvents. Trends Biotechnol 1990; 8:244-249.
12. Klibanov AM. Enzymes that work in organic solvents. Chem Tech 1986; 16: 354-359.
13. Zaks A. Protein-water interactions. In: Ahern TJ, Manning MC, eds. Stability of Protein Pharmaceuticals, Part A. Chemical and Physical Pathways of Protein Degradation. NY: Plenum Press Inc., 1992:249-271.
14. Ahern TJ, Klibanov AM. The mechanism of irreversible enzyme inactivation at 100°C. Science 1985; 228:1280-1284.
15. Mozhaev VV, Poltevsky KG, Slepnev VI et al. Homogeneous solutions of hydrophilic enzymes in nonpolar organic solvents. FEBS Lett 1991; 292:159-161.
16. Barone G, Del Vecchio P, Fessas D et al. Protein stability in non-aqueous media: a DSC study. In: van den Tweel WJJ, Harder A, Buitelaar RM, eds. Stability and Stabilization of Enzymes. Amsterdam: Elsevier BV, 1993:37-44.
17. Battistel E, Bianchi D. Influence of the solvent properties on protein stability in organic media. In: van den Tweel WJJ, Harder A, Buitelaar RM, eds. Stability and Stabilization of Enzymes. Amsterdam: Elsevier BV, 1993:13-20.
18. Zaks A, Klibanov AM. Enzymatic catalysis in organic media at 100°C. Science 1984; 224:1249-1251.
19. Hirata H, Higuchi K, Yamashina T. Lipase-catalyzed transesterification in organic solvent: effects of water and solvent, thermal stability and some applications. J Biotechnol 1990; 14:157-167.
20. Zaks A, Klibanov AM. Enzymatic catalysis in non-aqueous solvents. J Biol Chem 1988; 263:3194-3201.
21. Russell AJ, Klibanov AM. Inhibitor-induced enzyme activation in organic solvents. J Biol Chem 1988; 263:11624-11626
22. Volkin DB, Staubli A, Langer A et al. Enzyme thermoinactivation in anhydrous organic solvents. Biotechnol Bioeng 1991; 37:843-853.
23. Gorman LA, Dordick JS. Organic solvents strip water off enzymes. Biotechnol Bioeng 1992; 39:392-397.
24. Mozhaev VV, Khmelnitsky YuL, Sergeeva MV et al. Catalytic activity and denaturation of enzymes in water/organic cosolvent mixtures. Eur J Biochem 1989; 184:597-602.
25. Reslow M, Adlercreutz P, Mattiasson B. Organic solvents for bioorganic synthesis. Appl Microbiol Biotechnol 1987; 26:1-8.
26. Khmelnitsky YuL, Mozhaev VV, Belova AB et al. Denaturation capacity: a new quantitative criterion for selection of organic solvents as reaction media in biocatalysis. Eur J Biochem 1991; 198: 31-41.
27. Gupta MN. Enzyme function in organic solvents. Eur J Biochem 1992; 203:25-32.
28. Zaks A, Klibanov AM. The effect of water on enzyme action in organic media. J Biol Chem 1988; 263:8017-8021.
29. Wheeler CJ, Croteau R. Terpene cyclase catalysis in organic solvent/minimal water media: demonstration and optimization of (+)-alpha-pinene cyclase activity. Arch Biochem Biophys 1986; 248: 429-434.

30. Kazandjian RZ, Klibanov AM. Regioselective oxidation of phenols catalyzed by polyphenol oxidase in chloroform. J Am Chem Soc 1985; 107:5448-5450.
31. Mozhaev VV, Sergeeva MV, Belova AB et al. Multipoint attachment to a support protects enzyme from inactivation by organic solvents: alpha-chymotrypsin in aqueous solutions of alcohols and diols. Biotechnol Bioeng 1990; 35:653-659.
32. Wong C-H, Chen S-T, Hennen WJ et al. Enzymes in organic synthesis: use of subtilisin and a highly stable mutant derived from multiple site-specific mutations. J Am Chem Soc 1990; 112:945-953
33. Ziyang Zhong, Liu JL-C, Dinterman LM et al. Engineering subtilisin for reaction in dimethylformamide. J Am Chem Soc 1991; 113:683-684.
34. Ayala G, de Gomez-Puyou MT, de Gomez-Puyou A. et al. Thermostability of membrane enzymes in organic solvents. FEBS Lett 1986; 203:41-43.
35. Khmelnitsky Y, Welch SH, Clark DS et al. Salts dramatically enhance activity of enzymes suspended in organic solvents. J Am Chem Soc 1994; 116:2647-2648.
36. Riva S, Chopineau J, Kieboom APG et al. Protease-catalyzed regioselective esterification of sugars and related compounds in anhydrous dimethylformamide. J Amer Chem Soc 1988; 110:584-589.
37. Rose K, Vilaseca LA, Werlen R et al. Preparation of well-defined protein conjugates using enzyme-assisted reverse proteolysis. Bioconjugate Chem 1991; 2:154-159.
38. Schellenberger V, Jakubke H-D. protease-catalyzed kinetically controlled peptide synthesis. Angew Chem Int Ed Engl 1991; 30:1437-1449.
39. Murphy A, Ó Fágáin C. Bypassing the ribosome: peptide synthesis without translation. Essays Biochem 1996; 31:61-75.
40. Jakubke H-D. Enzymatic peptide synthesis. In: The Peptides 9. NY: Academic Press, 1987:103-165.
41. Zaks A, Klibanov AM. Enzyme-catalyzed processes in organic solvents. Proc Nat Acad Sci USA 1985; 82:3192-3196.
42. Dordick J.S. Enzymatic and chemoenzymatic approaches to polymer synthesis. Trends Biotechnol 1992; 10:287-293.
43. Zaks A, Russell AJ. Enzymes in organic solvents: properties and applications. J Biotechnol 1989; 8:259-270.

CHAPTER 4

Thermophilic Proteins

Introduction

There exists a wide variety of microorganisms capable of living at elevated temperatures. These include both eubacteria and archaebacteria. These organisms are of great interest in their own right and have been the subject of numerous monographs[1-6] and reviews.[7,8] Thermophilic organisms are generally defined as those capable of growth at temperatures ranging from 55-100°C.[6] Kristjansson has proposed that the term 'thermophile' be restricted to organisms with an optimum growth temperature in the range 65-85°C and that organisms with an optimum growth temperature above 85°C be referred to as 'extreme thermophiles'.[8] In such a scheme, bacteria growing optimally at 60°C or less would be described as 'slight thermophiles'. Brock has tabulated a number of thermophilic bacteria[7] which are available from the German Collection of Microorganisms (DSM, Göttingen).

Protein scientists are greatly interested in extremophile microorganisms. These comprise Archaea, eubacteria and some fungi. Many live at temperatures, pH values or salt concentrations that challenge protein integrity. The various types of thermophilic eubacteria are the best characterized[6,7,9,10] but the Archaea are receiving increasing attention.[9,11-13] Most hyperthermophiles (organisms growing in the range 80-110°C and incapable of growth below 60°C) are Archaea.[10] In thermophilic organisms, key proteins are fully able to tolerate high temperatures even following isolation from the microbial cell.

The organism *Thermus aquaticus*, isolated by Brock and Freeze in 1969,[14] has provided a DNA polymerase (Taq polymerase)[15] widely used in molecular biology for the polymerase chain reaction (PCR) for amplification of DNA copy numbers.[16,17] This thermostable DNA polymerase molecule can withstand the successive cycles of heating and cooling required to separate DNA duplexes and allow successive rounds of DNA replication to take place with coincident amplification of copy number. Indeed, nowadays it seems as if every supplier of molecular biology reagents offers heat-stable enzymes as tools for recombinant DNA work. Examples include the Vent™ and Deep Vent™ DNA polymerases from New England Biolabs (Beverley, MA) and the *Taq* and *Pfu* DNA polymerases and *Pfu* and *Tth* DNA ligases from Stratagene (La Jolla, CA). Boehringer Mannheim (Penzberg, Germany) supplies *Pwo, Taq* and *Tth* DNA polymerases. For biocatalytic applications, Recombinant BioCatalysis (Sharon Hill, PA) offers lyophilized cloned enzymes from extremophile microorganisms to aid in biocatalyst screening for processes. (Note that mention of company or product names does not imply approval or endorsement of any particular product. These examples are given merely to highlight the fact that enzymes from thermophilic organisms have already become commercially

Stabilizing Protein Function, by Ciarán Ó Fágáin.
© Springer - Verlag and Landes Bioscience 1997.

important.) Thermotolerant enzymes are already used in applications such as starch processing (alpha-amylase and pullulanase; ref. 18 and later in this chapter). Kristjansson has listed a number of other commercially available thermostable enzymes.[8] Considerable research has taken place in recent years on the genera *Thermus* and *Bacillus*. These are quite easy to cultivate and one can isolate their enzymes by conventional techniques. There has been progress also with molecular biological studies. Genes from some thermophiles have been cloned into mesophilic hosts and expression has been achieved in many cases.[19] Efficient gene transfer is possible in *Thermus thermophilus* and plasmids are available for genetic manipulation of *Bacillus stearothermophilus*.[7] Ward and Moo-Young have summarized examples of thermophilic enzyme expression in mesophilic hosts.[20]

Proteins derived from thermophilic organisms have been intensively studied and compared with their mesophilic counterparts in efforts to elucidate the molecular basis of their remarkable thermostabilities, as summarized in this chapter; see also refs. 21, 22. Voordouw and colleagues have proposed a definition of a thermophilic *enzyme* as one possessing a total kinetic thermostability ($\Delta G^{\ddagger}_{T}$, the total Gibbs energy of activation of the denaturing/inactivating process) greater than, or equal to, 25.0 kcal mol^{-1} at 70°C under conditions where the observed total kinetic thermostability is at a maximum.[23]

Thermophilic proteins show greater tolerance of high temperatures (but not necessarily of chemical denaturants; see ref. 24) than do their mesophilic counterparts. This enhanced heat stability is achieved by quite minor and subtle changes. The three-dimensional structures of thermophilic and mesophilic forms of a particular protein resemble each other closely.[25,26] If the two types of protein share similar tertiary structures, the difference in their stabilities must be due to changes in lower orders of structure, particularly primary structure. For instance, the tertiary structures of pig heart and (Archaeal) *Thermoplasma acidophilum* citrate synthases have a high degree of structural homology but only 20% sequence identity.[27]

Amino Acid Compositions

Amino acid sequence comparisons have been undertaken for a number of proteins to ascertain if any 'thermostabilizing' amino acid usage or substitution patterns occur. In 1975 Perutz and Raidt examined the structures of ferredoxins from mesophiles and thermophiles.[28] Ferredoxins are single chain polypeptides of less than 60 residues. Much information is available concerning their structures, heat stabilities and amino acid sequences. Amino acid substitutions leading to increased thermostability resulted in formation of additional external salt bridges. These had the effect of linking residues near the N-terminus to others near the C-terminus. For instance, ferredoxins from mesophiles contain an Ile residue at position 29 and an Asp at position 6. In a 'slightly thermophilic' species, Arg occurred at position 29 while Glu occupied position 6. This allowed a stabilizing Arg-Glu salt bridge to form. Similarly, salt bridges were observed between Glu and Lys residues which substituted in the thermophilic ferredoxin for Ser 7 and Val 53, respectively. In the thermophilic proteins, a His replaces Tyr 2 and seems to point towards a conserved Cys residue. Formation of an additional hydrogen bond between these two residues was postulated. The thermostable ferredoxins had a further hydrogen bond between Ser 10 and Tyr 30 (see Table 4.1 for a summary). Additional non-polar contacts did not appear to be formed.[28]

Table 4.1. Amino acid changes leading to increased ferredoxin thermostability.

Increasing thermal stability →	Why stabilizing
D6 → E	Formation of E6-R29
I29 → R	salt bridge
S7 → E	Formation of E7-K53
V53 → K	salt bridge
Y2 → H	Extra H2-C43 hydrogen bond
S10-Y30	Extra hydrogen bond in thermophilic ferredoxins

Block capitals represent the single-letter amino acid code while numerals denote position from N-terminus in the ferredoxin sequence. See reference 28. Reprinted from Biochim Biophys Acta, 1252 Ó Fágáin C, Understanding and increasing protein stability, 1-14, 1995 with kind permission of Elsevier Science - NL, Sara Burgerhartstraat 25, 1055 KV Amsterdam, The Netherlands. © 1995 Elsevier Science - NL.

Some years later Zuber adopted a similar but more rigorous approach to comparative studies of the lactate dehydrogenase (LDH) enzyme isolated from psychrophilic (cold-loving), mesophilic and thermophilic *Bacillus* strains.[29] These were *B. psychrosaccharolyticus* (psychrophile), *B. subtilis* and *B. megaterium* (mesophiles) and *B. stearothermophilus*, *B. caldotenax* and *B. caldolyticus* (thermophiles). These enzymes show 60-70% sequence similarity and much structural and functional information is available, including three-dimensional structures for the analogous muscle enzyme. Thermophilic LDH seemed to have more hydrophobic interactions and ion pairs. Phe and Val residues tended to substitute in the more thermophilic strains for Ile and Leu while Ala and Met residues occurred among the more mesophilic bacteria studied. These 'thermophilizing' substitutions lead to increased hydrophobicity and packing density in the LDH molecule. Among charged residues, Arg and Asp seem to be more thermostabilizing (perhaps functioning as ion pairs) than do Glu and Lys. (Arg has a higher pK than Lys: see next paragraph. The preference for Asp is perhaps surprising in view of its chemical lability, as outlined in chapter 2.) These sorts of amino acids could participate equally well in hydrogen bonding with polar residues or with peptide bonds. However, one must also consider steric and energetic considerations in explaining these replacements. Zuber also raises and discusses the interesting idea that the thermostabilizing amino acid substitutions may result from temperature-related base substitutions in mRNA translation.[29]

Arginine seems generally to substitute for lysine among thermophilic proteins.[25,29] As noted by Volkin and Middaugh,[25] arginine's pK_a (approx. 12) is higher than that of lysine (approx. 9.5). Arginine remains positively charged at highly alkaline pH values and also remains charged at higher temperatures (compared with lysine) since pK_a decreases with increasing temperature. The distance from the polypeptide backbone to the positive charge is shorter by one $-CH_2-$ group in arginine than in lysine, reducing any unfavorable hydrophobic contacts with solvent. The bulky guanidino function in arginine gives a greater positively-charged surface for stabilizing interactions than does the positive amino function of lysine. Lastly, lysine is prone to the destructive Maillard reaction with reducing sugars.[25] All of these features favor arginine over lysine as a heat-stable or stabilizing positively charged residue.

Other 'thermophilizing' substitutions have been noted. Phenylalanine and valine tend to substitute in thermophilic *Bacillus* lactate dehydrogenases (LDH) for the 'mesophilic' isoleucine and leucine, leading to increasing hydrophobicity and packing density in the LDH molecule.[29] Sequence studies on glyceraldehyde 3-phosphate dehydrogenases and 3-phosphoglycerate kinases from various sources suggest that polypeptide adaptations to higher temperatures involve increases in average hydrophobicity and decreases in chain flexibility.[27]

Suzuki has likewise studied a number of Bacillus species, focusing on the enzyme oligo-1,6-glucosidase.[30] As thermostability increased in the order *B. cereus* → *B. coagulans* → *B. thermoamyloliquefaciens* → *B. thermoglucosidasius* → *B. flavocaldarius*, the proline content of the respective enzymes increased in a linear fashion. The potential for hydrophobic interactions was also observed to increase with greater thermostability. Proline's R-group is a five-membered imino ring which is very rigid and leads to kinks or bends in a folded protein. This greatly restricts the conformational flexibility of the protein backbone and tends to rigidify the molecule against unfolding. Proline was noted to occur at the second site of a beta-turn, helping to reduce the entropy of unfolding.[30] Suzuki suggests that such substitutions may be useful as a general principle of protein stabilization. These observations coincide with those of Matthews and colleagues who noted a small increase in the denaturation temperature of lysozyme from bacteriophage T4 when site-directed mutagenesis substituted Pro for the wild type Ala at position 82 in a beta-turn.[31] Recall, however, from chapter 2 that cis-trans isomerization of Pro residues can occur, leading to disruption of polypeptide structure.[32]

More recently, Suzuki's group has reported on a series of *B. cereus* ATCC7064 oligo-1,6-glucosidase variants multiply substituted with Pro residues.[33] The *B. cereus* enzyme was the least stable among the six alpha-1,6-glucosidases studied in formulating the 'proline rule' in ref. 30 and in the introduction of prolines based on the *B. cereus* enzyme's three-dimensional structure[34] and the primary structures of thermostable counterparts.[35] The X-ray structure revealed that the enzyme comprises three domains. Sequence comparisons indicated that each enzyme contains 22 critical sites where proline may occur randomly with increasing stability. Most of these critical sites occupy the second position within solvent-exposed beta turns, while others occur within loops forming the active site cleft and its wall. Some of the critical sites comprise the N-caps of alpha helices. In the naturally thermostabilized variants, proline residues at the critical sites tend to replace charged or polar residues.[33] Nine positions (Lys121, Glu175, Glu290, Glu208, Glu270, Glu378, Thr261, Glu216 and Asn109) were cumulatively replaced by proline in the order listed. Most positions were exposed to solvent. Of these nine mutations, three (Lys121, Glu208 and Glu290) occupied the second site within a beta turn, three more (Asn109, Glu175 and Thr261) comprised the N-caps of alpha helices and the final three (Glu216, Glu270 and Glu378) occurred in coils within loops. Additive increases in thermostability were recorded with successive proline replacements. The half-inactivation temperature, T_{50}, increased from 44.5°C for the wild-type *B. cereus* enzyme to 49.6°C for the variant incorporating all nine substitutions. Curiously, a plateau was reached between 4 and 7 substitutions in terms of T_{50} values.

The cumulative replacements K121P, E175P, E290P and E208P all increased the T_{50} value. Incorporation of the further cumulative mutations E270P and E378P slightly decreased the T_{50}, but the final cumulative replacements T261P E216P and N109P again led to increases in T_{50}. The two 'anomalous' replacements, E270P and E378P, which failed to increase T_{50}, also yielded 48°C half-lives against the generally-stabilized trend. Both of these 'ineffective' substitutions occurred in coils, as did E216P. The most noteworthy stability enhancements occurred when Pro was inserted at the second sites of beta turns (K121P, E290P, E208P) or at the N-caps of alpha helices E175P, T261P, N109P). The results were interpreted as definitive supporting evidence for the 'proline rule' of protein thermostability enhancement.

Polar amino acids, such as serine or threonine, are less common in thermophilic proteins and their cysteine content may be minimal or nil (see, for instance, ref. 27). This in turn means that the naturally-occurring crosslink of the disulfide bond occurs to no greater extent than in mesophilic proteins. Although the heat lability of -S-S- bridges has been discussed in chapter 2, the relative absence of cysteine and -S-S- links may not in itself be an adaptation for high-temperature function. Prokaryotic enzymes in general have lower cysteine contents than eukaryotes and the scarcity of -S-S- links may reflect bacteria's lack of an endoplasmic reticulum for extensive post-translational modification of newly-synthesized proteins.[25] The Asn and Gln contents of thermophilic proteins are much reduced compared with their mesophilic counterparts[36] since these residues are prone to destructive deamidation (chapter 2). Whether this reduced Asn/Gln prevalence is a general 'thermophilizing' effect among the Archaea remains to be seen. Hensel and colleagues have shown that the native conformation of a polypeptide backbone can prevent deamidation of particular Asn and Gln residues by steric hindrance and restricted torsional freedom and that this protection is lost once the protein unfolds. They also suggest that Asn and Gln may be restricted in thermophilic proteins to such conformationally-protected locations.[24,36]

While the types of amino acid substitutions discussed above point towards the sorts of changes required for thermostability, it is not yet appropriate to postulate a widely applicable mechanism for thermophilic protein stability, since it is specific molecular interactions and forces that contribute to the stabilities of individual proteins. The net Gibbs (free) energy of stabilization of a protein is generally about 5-15 kcal/mol (21-63 $kJ.mol^{-1}$), quite a small value. The appropriate combination of the weak bonds involved in maintaining higher order structure in proteins can give rise to energy differences of this magnitude. Among all of the molecular interactions leading to thermostability, hydrophobic interactions seem to be particularly important.[37-39] Daniel et al argue that some of the predicted amino acid changes associated with thermophily (such as an increased Arg/Lys ratio) have little predictive value or statistical significance and that the Archaeal origin of some enzymes seems a more likely source of the observed variations in amino acid compositions. Taxonomic or functional replacements of amino acids in enzymes provide a high baseline of variation above which small and particular 'thermophilizing' substitutions cannot be statistically detected. Similarly, no general or systematic mechanism(s) for the achievement of thermostability have emerged to date from structural comparisons of homologous enzymes.[37]

Flexibility Within Protein Molecules

Note that thermophilic proteins must balance the conflicting demands of molecular flexibility (for efficient catalysis) and stability (which requires extra interactions).[36] Thermophilic enzymes appear to have somewhat less conformational flexibility and less catalytic efficiency (at least at mesophilic temperatures) than do their mesophilic counterparts.[38] Their greater molecular rigidity gives thermophilic proteins extra stability, but at a price: a thermophilic enzyme is less able to undergo the conformational flexions required for substrate torsion during catalysis *where the activities of both enzymes are measured at the same temperature.* When, however, each enzyme is placed at the temperature to which it has adapted by evolution, their molecular flexibilities are similar.[37] Recall that the thermophilic protein has developed to *function at* high temperatures, not just to tolerate them. In terms of relative catalytic effectiveness, it is difficult to compare validly the specific activities of similar enzymes from mesophiles and thermophiles. However, the available evidence suggests similar stabilities of mesophilic and thermophilic enzymes at similar levels of molecular flexibility.[37] For instance, phosphoglycerate kinases from thermophilic and mesophilic organisms had comparable catalytic efficiencies when assayed at their respective optimum temperatures.[40] Likewise, kinetic parameters of the *Thermoplasma acidophilum* citrate synthase recorded at its optimum temperature of 55°C closely resembled those of the pig heart enzyme at mesophilic temperatures. Also in citrate synthase, a substrate-binding flexible loop in the pig heart enzyme is absent from the *T. acidophilum* enzyme. This illustrates once again the balance required between flexibility for catalysis and a certain degree of rigidity necessary for stability,[27] a point made in the conclusion to chapter 2. See also Wrba and colleagues[41] who have proposed that thermophilic proteins are less flexible at mesophilic temperatures than are mesophilic proteins, but that both types of protein are similarly flexible at their respective optimum temperatures of activity.

Daniel et al have critically reviewed the high-temperature denaturation and degradation of enzymes. These workers have noted that, in recent studies of enzymes remaining functional *above* 100°C, catalytic activities have actually been measured at these extremely high temperatures. They go on to point out that simple exposure of an enzyme to temperatures >100°C, followed by cooling and assay of remaining activity at a lower temperature (e.g. 95°C) is not an ideal methodology. In such an experiment, the observed activity may arise not alone from thermostable, native enzyme molecules (N; see Fig. 1.3) completely resistant to unfolding but also to molecules which unfolded (U) at the higher temperatures but renatured rapidly (to N) upon cooling. In other words, one may be estimating the combined activities of N and refolded U enzyme forms.[37] (Similar concerns were addressed in a study of mesophilic recombinant horseradish peroxidase.[40]) While no explicit case of such a rapid renaturation has been reported, activity measurements at high temperatures greatly reduce the risk of this possible artifact.[39]

Daniel et al argue that the relative instability of mesophilic enzymes is a functional requirement, in order that they may have the requisite flexibility for efficient catalysis at lower mesophilic temperatures; also, 'excessively' stable proteins will be more resistant to proteolysis in vivo and hence more difficult to degrade during normal protein turnover within the cell. (It is important to bear in mind, however, that stability is a 'global' property of the entire protein, while a merely-local flexibility may be all that is required for effective catalysis or other function.[37])

Features of Some Thermophilic Proteins: Citrate Synthase

The crystal structure of citrate synthase from *Thermoplasma acidophilum* was reported in 1994.[27] This enzyme is a homodimer. The active site is situated between the small and large domain of each monomer and comprises amino acid residues from both monomers. The overall three-dimensional structures of the citrate synthases from *Thermoplasma acidophilum* and pig heart enzymes are very similar, but notably different in the loops connecting alpha-helical regions. Many of the loops occurring in the Archaeal enzyme are much smaller than those of the pig heart enzyme; indeed, four loops are absent from the Archaeal monomer. The highly flexible interhelical loops in pig heart citrate synthase may act as weak points during thermal unfolding while the shorter loops in the *T. acidophilum* enzyme may help minimize unfolding. The Archaeal enzyme is also much more compact than the mesophilic one. Calculations by two different methods revealed that the volume of cavities within the structure of the Archaeal enzyme is less than half of that within the mesophilic synthase *at room temperature*; the volume of cavities within the Archaeal structure at a higher temperature has yet to be ascertained. The closer packing results from a reduction in the number of helices and especially of loops in the Archaeal enzyme.[27] The Archaeal synthase is 35 amino acids shorter than the pig enzyme at its N-terminus. This shortening occurs also in citrate synthase from other extremophile Archaea and from a thermotolerant *Bacillus*. The N-terminal differences may relate to stability but this is not certain, as similar shortened citrate synthase N-termini have been noted in the mesophiles *Pseudomonas aeruginosa* and *E. coli*.[27]

Aldehyde Ferredoxin Oxidoreductase

In 1995, Chan et al described the crystal structure of aldehyde ferredoxin oxidoreductase (AOR) from the hyperthermophilic archaeon (archaebacterium) *Pyrococcus furiosus*, where it functions in the oxidation of aldehydes (aromatic or aliphatic) to carboxylic acids with an optimum temperature >95°C. This was the first reported structure of a hyperthermophilic enzyme. The polypeptide comprises two identical subunits; each subunit contains an Fe_4S_4 and a molybdopterin-coordinated tungsten cofactor. It is thought that two molybdopterin molecules ligate the tungsten with a total of four sulfur ligands. No protein ligands coordinate the tungsten. There are no disulfides. Aside from the iron-sulfur cluster and the tungsten atom, which lie close to each other within each subunit, there is a more distant single tetrahedral metal situated at the interface between the dimers. Because of its distance from the other metal atoms, this metal is more likely to stabilize the dimer at high temperatures than to have a redox function. No structure of a corresponding mesophilic protein is available for direct comparison, however. When compared generally with other known protein structures, hyperthermophilic AOR has quite a small solvent-exposed surface area. It is also distinguished by large numbers of ion pairs and buried atoms. The enzyme's notable thermostability may be due to all of these features.[43]

Thermolysin and Neutral Protease

Fontana has studied diverse thermophilic proteins and especially thermolysin. This is a zinc-dependent, calcium-containing endopeptidase from *Bacillus thermoproteolyticus*. It is a single chain polypeptide lacking disulfide bonds and was the first thermophilic protein to be determined by X-ray methods.[38] Thermolysin appears to undergo selective peptide bond fissions when inactivated by limited proteolysis, heat autolysis or ion chelation by EDTA. Such cleavages allow isolation of 'nicked' thermolysin fragments and these have been correlated with the known three-dimensional structure. The cleavages are all found at exposed loops, never within an ordered secondary structure. Some of the fragments generated by these fissions correspond to domains or subdomains of the native protein and are capable of folding to native-like structures in aqueous solution.[38]

The amino acid sequence of thermolysin differs from that of the homologous *B. stearothermophilus* neutral protease (NP-ste) in 45 positions. Thermolysin is 13 degrees more thermostable than NP-ste. Eijsink and colleagues undertook site-directed mutagenesis of NP-ste to ascertain which of these amino acid differences had a bearing on stability. The NP-ste residues at these different positions were replaced with the corresponding thermolysin amino acid. For comparison, cavity-filling and cavity-making mutations were designed into the hydrophobic protein core. (For this purpose, a three-dimensional model structure of NP-ste was devised, based on the known thermolysin crystal structure.) Also, a region critical to stability which was identified by the thermolysin-substitutions approach was rigidified by Ala→Pro substitutions. The index of stability used was T_{50}, the temperature at which 50% caseinolytic activity remained following a 30-minute incubation at pH 5.2. (Except for the slightly-stabilizing G141A, each mutant's specific activity closely resembled that of the wild type.) Changes in T_{50} were used to quantify the effects of mutations. It turned out that just a few of the sequence differences accounted for the total difference in stability between the two enzymes and that all of these residues lay on the protein surface. Positions 63-69 seemed to be particularly important: substitutions at residues 63 and 69 were markedly stabilizing. For example, the T63F and A69P substitutions led to T_{50} increases of 6.0° and 5.6°C, respectively, and were three times more stabilizing than any of the other favorable surface mutations. In contrast, mutations within the hydrophobic core had little effect: none of those studied influenced T_{50} by more than 0.4°C either way. From these results, Eijsink et al inferred that the thermoinactivation rate is a function of local unfolding processes (at the protein surface) which act as foci for autolysis. The stabilizing mutations within the 63-69 region are thought to prevent the initial events of the unfolding process. In an attempt to decrease the NP-ste molecule's flexibility in this crucial region, proline substitutions were made at positions 63, 65, 66 and 69. The mutants S65P and A69P increased T_{50} by 4.2° and 5.6°C, respectively but the Pro substitutions at positions 63 and 66 were destabilizing. The authors argue that mutations within an unstable (early-unfolding) region of a protein are more beneficial than alterations within a relatively stable region which would be less prone to autolysis.[44]

Thermophilic Amylases

Microbial amylases are a very intensively studied group of enzymes. Progress in this area during the 1980s has been reviewed by Vihinen and Mantsala,[45] who described the properties of a variety of thermophilic amylases from *Bacilli* and from a smaller number of actinomycetes.

Many factors can affect thermostability of amylases, including the presence of calcium and of substrate. The bound starch substrate is thought to rigidify the enzyme's conformation and most starch preparations contain some calcium as an impurity. (Some thermoactinomycete amylases, however, are not stabilized by starch.) Calcium has an activating and stabilizing effect on alpha-amylase. The bound calcium may be responsible for tightening the binding between the different domains of the enzyme. (It has been suggested that many experiments may not in fact have measured true thermostabilities due to the use of too short an incubation period.[45]) Vihinen and colleagues described a His238→Asp mutant of *B. stearothermophilus* alpha-amylase which was less thermostable than the wild type protein. The mutation was thought to cause a more open, flexible backbone structure, leading to the observed loss of stability.[46] Many organisms growing at mesophilic temperatures produce alpha-amylases maximally active at higher temperatures. The mesophilic *B. licheniformis* produces amylases with temperature optima in the range 70-90°C. The enzyme from *B.licheniformis* MB 80 can be used satisfactorily at 100°C.[47]

Temperature-shift selection techniques have been applied to 'improve' the thermophilic alpha-amylase from *Bacillus licheniformis*. Joyet and co-workers reported the isolation of a hyperthermostable variant of this enzyme following temperature shift and in vivo screening.[48] Their strategy involved hydroxylamine mutagenesis of a *thermolabile* alpha-amylase from a mutant *B. licheniformis* strain [D164A-W165G] expressed in M13 phage. Selection allowed the isolation of an intragenic suppressor mutation which overcame the initial, thermolabilizing mutation. The stabilizing suppressor mutation (i.e. which restored stability to the heat-sensitive, initial amylase mutant) was subsequently introduced into the normal, wild type alpha-amylase, where it should confer additional heat stability. While this protocol may seem involved, it was necessitated by the difficulties of selecting for a more stable form of an already thermostable protein. Inactivating temperatures for the alpha-amylase are too high for growth of typical host bacteria or even of thermophilic organisms. Also, an amylase enzyme with enhanced thermotolerance will give no selective advantage to its host cell. Suppressor mutations revealed by the selection process contained only a single amino acid change, V209A; no revertants to the wild type sequence were observed. The 90°C half life of an alpha-amylase incorporating this mutation was 2-3 times greater than that of the wild type enzyme. The combined effects of the V209A mutation and a previously-discovered H133Y replacement were additive: the double mutant's 90°C half life was 12 hours, 9-10 times longer than those of the wild type or an industrial thermostable alpha amylase ('Termamyl'®; see Table 4.2). Both mutations are situated in a folding domain, the B domain, which apparently is critical for the enzyme's stability. Joyet et al postulate that the B domain forms a weak link in the protein where unfolding is initiated. The A209V substitution perhaps strengthens hydrophobic interactions in the B-domain (as the H133Y replacement is known to do; see ref. 49), preventing unfolding. Random mutagenesis was the only viable pro-

Table 4.2. Thermal stability parameters for Termamyl® and* B. licheniformis *H133Y-A209V double mutant alpha-amylase.

	T_{50} (0.5)	T_{50} (2)	T_{opt}	$t_{1/2}$ (90°C)	Ca^{2+}_{50} (mM)
Termamyl®	107°C	114°C	90°C	130 min	0.50
H133Y-A209V	118°C	125°C	95°C	>12 h	0.24

T_{50} signifies the temperature at which 50% inactivation occurs in 15 min at the millimolar calcium concentrations indicated in brackets. T_{opt} is the temperature optimum for activity in the presence of 0.5% soluble starch at pH 6 with 1mM calcium. $t_{1/2}$ denotes half life, measured at pH 6 with 1mM calcium. Ca^{2+}_{50} signifies the concentration of calcium ion at which the enzymes express half their maximal activity. Data from ref. 48.

tein engineering strategy for the *B. licheniformis* alpha-amylase, the crystal structure of which was not then available. The double occurrence of the same suppressive mutation -A209V- suggests that there are very few second site mutations in the alpha-amylase gene capable of overcoming the initial, heat-sensitizing mutation. Note, however, that hydroxylamine mutation cannot 'sample' all possible codons; thus, only a few amino acid substitutions were possible. (See conclusion of chapter 8 for a further explanation of this point.) Val, a more hydrophobic residue than Ala, occurs at position 209 in the alpha-amylases from *B. amyloliquefaciens* and *B. stearothermophilus*, but both of these enzymes are less stable than the wild type *B. licheniformis* enzyme. Accordingly, this beneficial mutation could not have been predicted from sequence comparisons.[48]

The strict anaerobe *Dictyoglomus thermophilum*, which grows optimally at 70°C, produces multiple extracellular amylases. Three of its amylase genes have been successfully cloned and expressed in *E. coli*.[50]

Thermophilic alpha-amylases have proven valuable in starch processing. They can begin hydrolysis of alpha-1,4 glycosidic linkages within the starch substrate at the high temperatures (90-105°C) required for dispersion or gelatinization of starch granules. alpha-Amylases, however, have a significant drawback, namely their inability to cleave alpha-1,6 glycosidic linkages occurring at branch points within the starch polymer. This leads to formation of partially hydrolyzed starch limit dextrins which cannot be processed further. Fortunately, this problem can be relieved by the use of microbial pullulanases (EC 3.2.1.41). (Pullulan is an alpha-1,6 linked linear polymer of maltotriose subunits.) These enzymes have been successfully used as debranching catalysts in starch hydrolysis, but to date most have come from mesophiles and must be used at or below 60°C. A heat-stable pullulanase would obviously be superior and recent studies have shown pullulanase activities in thermophilic *Clostridium* and *Thermus* species.[51] Starch-limited culture conditions led to good pullulanase production levels for the *Clostridia* studied and the enzyme from *C. thermohydrosulfuricum* is stable up to 90°C.[51,52] This, and many other pullulanases, can, in addition, act on the alpha-1,4 linkages within starch: it is not clear why this is so.[51] One should note that the slightly acid pH optima (5-6) of the thermophilic pullulanases resemble those of the glucoamylase and beta-amylase enzymes used in the later (saccharification) stages of starch processing. Alpha-amylases generally have their pH optima at pH 6.5 and require calcium ions, unlike pullulanases. These differences could result in redesigned processes for starch hydrolysis.[51]

Thermophilic Glucose Isomerases

Microbial xylose isomerases can often accept glucose as an alternative substrate, converting glucose to its sweeter-tasting keto isomer, fructose. Xylose/glucose isomerases (GIs) are widely used for the production of high-fructose syrups for the food and soft drink industries. (See also chapters 7 and 8.) There are two classes of GI. Class I enzymes have shorter N-terminal regions than their Class II counterparts. Examples of Class I enzymes are those from *Arthrobacter*, *Streptomyces* and *Actinoplanes missouriensis*. Class II includes the GIs from *E. coli*, *Bacillus* and *Thermoanaerobacterium*.[53] Higher process temperatures lead to higher fructose yields.[54] Accordingly, there is great interest in the isolation and use of more heat-tolerant GI enzymes to aid process efficiency.

The GI enzyme from *Thermus thermophilus* is encoded by a single gene. Each of four identical subunits has a molecular weight of 44,000 and comprises 387 amino acids. Like most GIs, it functions as a homomeric tetramer with a molecular weight of 200,000. Cloning of the *Thermus thermophilus* GI gene into *E. coli* with use of the *tac* promoter led to a 45-fold increase in GI yield compared with production in the natural host. Also, production in *E. coli* affords a simple purification regime: a 10-minute, 85°C heat step purifies GI 20-fold from *E. coli* cell extract. Comparison of the amino acid sequence of *T. thermophilus* GI with those of other thermotolerant GI enzymes (from *Streptomyces violaceoniger, Actinoplanes missouriensis, Ampullariella* sp., *Clostridia, B. subtilis* and *E. coli*) indicates that the increased thermostability of the *T. thermophilus* enzyme arises from a decreased prevalence of thermolabile amino acid residues and from increased hydrophobicity. A number of Gly-to-Pro substitutions are noticeable in the *T. thermophilus* GI. Preferred amino acids in the *T. thermophilus* composition are (in order of preference) Glu, Arg, Tyr, Val, Pro and Met, Leu, Cys, Asn, Lys and Trp. Disfavored amino acids are (least favored first) Asp, Ile, Gly, Thr, Gln, Ala, His, Ser and Phe. The most frequent exchanges of least-favored for most-preferred amino acids are Asp→Glu, Lys → Arg, Ile→Leu→Tyr, Ile or Thr→Val, Gly→Pro, Gly→Ala and Gln→His. Many of these replacements likely increase hydrophobicity and internal packing, while the Gly→Pro changes rigidify the polypeptide backbone. Substitutions of these sorts will tend to prevent unfolding. However, the decreased prevalence of 'thermolabilizing' residues such as Asp, Gln and Gly suggest that decreased rates of irreversible inactivation may also be a feature of this enzyme.[54]

A thermostable GI from a novel thermophilic *Streptomyces* sp. (PLC) strain, isolated from Nigerian compost soil, has been described. The enzyme was purified 9.3-fold from a crude cell extract with a 10% yield. The native homotetramer has a molecular mass of 183 kDa; its subunit molecular mass is 46 kDa. In each subunit, two metal ions bind to non-identical sites and full catalytic activity is retained up to 95°C in presence of 0.4 M $MnSO_4$. Inactivation kinetics at 98°C were first-order with a half life of 15 minutes. At 53°C, the measured half lives in the presence of Co^{2+} and Mg^{2+} were 7 and 9 days, respectively. With Mn^{2+} as metal ion, activity remained constant for 10 days at this same temperature. Catalytic and other properties of this new GI are also detailed. The new strain is easy to grow and one can achieve large cell yields with high GI activities.[55]

Another thermophilic Class II GI, this time from *Thermotoga neapolitana*, has been cloned, sequenced and characterized by Vieille et al. The optimal temperature and pH are 97°C and 7.1, respectively while its calorimetric melting point is 88°C. Kinetic parameters of the *T. neapolitana* GI in the conversion of glucose to fructose were: k_{cat} 1,139 min^{-1}, K_m 88.5 mM. These values compare very favorably with other Class II GIs and even with some Class I counterparts (although the latter are much less thermostable). *T. neapolitana* GI retains significant levels of activity at pH values <7. This could permit the in-process use of acidic pH values to help minimize the Maillard reaction between glucose and the amino groups of lysine residues and the N-terminus. As with other thermophilic proteins,[36] when amino acid compositions and sequences are compared with those of other Class II GIs, the number of asparagine and glutamine residues in the GIs decreased as thermostability increased. This, presumably, is an adaptation aimed at minimizing deamidation reactions at high temperatures.[56] The enzyme is 70% identical to the GI from *T. thermosulfurigenes* 4B.

Meng and associates substituted some of the active site residues in *Thermoanaerobacterium thermosulfurigenes* GI by site-specific mutagenesis, thereby increasing its intrinsic thermostability.[57] (The uses of site-specific mutagenesis as a tool for the investigation of protein stability have been outlined in chapter 2. The purposeful use of specific mutations to enhance stability is considered in detail in chapter 8. Since the present example concerns a thermophilic GI, it seemed more appropriate to deal with here rather than in another chapter.) In common with other GIs, the *T. thermosulfurigenes* enzyme requires divalent cations (Co^{2+}, Mn^{2+} and Mg^{2+}) for catalytic activity and for themostability. Meng and colleagues focused on aromatic residues in the active site of the *T. thermosulfurigenes* enzyme and their substitutions had a dual purpose: to ascertain whether a reduced water-accessible hydrophobic surface area could increase the GI's thermostability, and whether thermostability is influenced by the hydrophobic interaction between aromatic residues which holds the monomers together as an active dimer.[57] The amino acids concerned are Trp49, Phe60, Trp139, Phe145 and Trp188. Like Dekker et al,[54] Meng's group exploited the relative thermostability of the *T. thermosulfurigenes* GI as a convenient means of purification from the *E. coli* cells into which the GI gene had been cloned. Trp188→His and Phe145→Lys GI mutants no longer resisted this heat step, indicating that the hydrophobic nature of these two residues are critical for thermostability. Mutations at Trp49 had differing effects. Trp49→Ala and Trp49→Phe were much less stable than wild type GI but Trp49→Arg had a longer half life than wild type. Replacement of Trp139 with smaller hydrophobic residues (Phe, Met and Ala) also led to gains in half life; the bulky indole side chain of Trp negatively affects the GI's stability. Replacement of Trp139 with smaller residues decreases the hydrophobic surface area within the active site and stabilizes the enzyme. Results also showed that 100 mM K^+ prolonged half life by seven times at 88°C and suggested that at least one potassium binding site exists in the *T. thermosulfurigenes* GI; the potassium is thought to stabilize the conformation and to prolong the half life. (Some thermophilic archaebacteria contain high intracellular K^+ concentrations: see ref. 36 and chapter 5.) Persistence of catalytic activity was estimated by following the time course of irreversible inactivation at 85°C with measurement of remaining activity at 65°C. Half lives were estimated from the first-order rate constants of thermoinactivation. It appeared from the experimental data that the formation of incorrect (or 'scrambled') structures is the main process leading to inactivation at elevated temperatures. Inactivation rates did not depend on protein concentration, indicating that some monomolecular event, rather than a polymolecular process such as aggregation, was responsible for the observed loss of activity.

Other Examples

Merkler and colleagues have examined the enzyme glutamine synthetase from the extreme thermophile *Bacillus caldolyticus*.[58] Two regulatory isoforms of this enzyme exist as separate gene products. Both of these forms have been observed to undergo a two-state aggregation to higher molecular weight species. The first stage involves polymerization of native dodecamers and this is followed by an aggregation process that results in the formation of amorphous, denatured protein. The initial aggregation was noted to occur at temperatures less than those required for time-dependent denaturation. The corresponding mesophilic *B. subtilis* enzyme seemed not to undergo any such temperature-dependent aggregation. It was suggested that aggregation may act as a stabilizing mechanism in this thermophilic system. Various metal ions and substrate molecules bind to the glutamine synthetase enzyme and bring about striking increases in the temperatures at which aggregation and activity loss occur. The authors hypothesized that denaturation results from unfolding of protein subunits within the dodecameric structure, rather than from unfolding to monomers before aggregation.

Aspartate aminotransferase has been purified to homogeneity from the thermophilic archaebacterium *Sulpholobus solphataricus*. Optimal growth conditions for this organism are 87°C at pH 3.5.[59] This enzyme is a homodimer of subunit molecular weight 53,000 and shows two main bands of pI 6.3 and 6.8 on isoelectric focusing. Temperature optimum was greater than 95°C. A 22-residue N-terminal sequence showed no similarity with any length of animal or bacterial aspartate aminotransferase sequence. This is remarkable because, at the level of the overall sequence, aspartate aminotransferases from cytosol, mitochondrion and *E. coli* are quite closely related.[59] Interestingly, the *S. solphataricus* enzyme has only 2 mol histidine per mol subunit; it also has an unusually high lysine content (43 mol per mol subunit) and a correspondingly lesser arginine content. This is in marked contrast to the idea expressed by Zuber[29] above. The Arrhenius plot for the *S. solphataricus* enzyme shows a break around 58°C. This indicates the occurrence of two different activation energy-dependent processes and has been interpreted as a thermal conformational transition.

Fabry and Hensel have reported the occurrence of a conformational change at 52°C in the 3-phosphoglyceraldehyde dehydrogenase enzyme (a homotetramer) from *Methanothermus fervidus*, as indicated by breaks in Arrhenius and Van't Hoff plots. The enzyme is stable up to 75°C and up to 90°C in the presence of $NADP^+$ at high ionic strength.[60]

Amino acid dehydrogenases are NAD(P) linked enzymes involved in reversible deamination reactions of L-amino acids. They have an alkaline pH optimum and are useful in the preparation of some L-amino acids for foodstuffs and nutritional preparations and in diagnostic assays.[61] The leucine dehydrogenase (LeuDH) from *Bacilli* is a hexamer of six or eight identical subunits weighing 245-310 kDa. The enzymes from the *B. sphaericus* and *B. stearothermophilus* (moderate thermophile) are very similar except for their stabilities. The *B. stearothermophilus* enzyme's optimum temperature is 72°C, roughly 15° higher than that of its mesophilic counterpart. The thermophilic LeuDH tolerates 60 minutes' incubation at 70°C and a wide range of pH. Unlike the mesophilic LeuDH, the *B. stearothermophilus* enzyme is stable on storage at 6°C for one month. It is also more resistant to detergent and ethanol denaturation.[61] The gene encoding *B. stearothermophilus* LeuDH has been cloned into *E. coli*, where the expressed enzyme can amount to 3% of the total soluble protein and from which it can be purified using its thermostability.[62] A 70°C heat step at pH 5.4 lasting 30 minutes yielded a six-fold purification of LeuDH with 93% activity recovery. Use of thermostable LeuDH produced in this way led to greatly improved L-leucine production in a membrane-based bioreactor system due to the longer enzyme half life.[63] A similar cloning and purification strategy has proven useful for the production of thermostable alanine dehydrogenase from *B. stearothermophilus*.[61]

There are many other reports describing notable properties of enzymes from thermophilic sources. These include investigations on lactate dehydrogenase from *Thermotoga maritima*, an extremely thermophilic eubacterium.[64] The enzyme shows little accessibility to proteases and is unusually resistant to guanidine hydrochloride. Compared with mesophilic enzymes, hydrogen-deuterium exchange rates are much reduced. The enzyme is a homotetramer weighing 140 kDa. Unlike eukaryotes, its N-terminus is not acetylated but its 30 N-terminal residues show a high degree of similarity to those of *Thermus aquaticus*. The enzyme is stable up to 90°C and NAD^+ enhances stability.

The widely-studied fungus *Aspergillus niger* has been shown to produce thermostable extracellular alpha- and beta-galactosidases. pH optima for both enzymes lay in the range 3.5-4.0 while each had a temperature optimum of 65°C. The alpha-galactosidase displayed a half-life of 104 minutes at 60°C while the beta-galactosidase's half-life at the same temperature was 835 minutes.[65] Both enzymes may eventually prove useful to industry, since alpha-D-galactosyl oligosaccharides occurring in soya beans cannot easily be digested by humans and since the beta-galactosidases used to date for lactose hydrolysis in milk and whey have all been thermolabile. It should be mentioned here also that Pisani et al have described a thermostable beta-galactosidase isolated from the thermophilic archaebacterium *Sulpholobus solphataricus*. The enzyme was purified to homogeneity and comprised a tetramer of 240 kDa. Serum raised against the enzyme did not cross-react with the corresponding *E. coli* enzyme. The *Sulpholobus* enzyme had a lower cysteine content and a lower arginine/lysine ratio than its *E. coli* counterpart.[66]

Enzymes with unusually high thermostability sometimes occur in non-thermophilic organisms, as evidenced by the malate dehydrogenase of *Rhodospirillum rubrum*, a non-sulphur purple phototrophic bacterium. This protein is capable of withstanding a temperature of 65°C for over one hour without any activity loss.[67] The corresponding enzyme from related phototrophic bacteria lost activity at much lower temperatures. Curiously, the *R. rubrum* enzyme had an amino acid composition very similar to those of the other thermophilic malate dehydrogenases and was in fact the richest in cysteine, often thought to be a 'thermolabile' amino acid residue. Such observations underline the extreme subtlety of some 'thermostabilizing' changes in proteins, a point developed more fully by Fontana[38] and by Jaenicke.[68]

Future Applications

Enzymes from thermophilic organisms may have particular usefulness in biosensor configurations, where long-term reliability is a critical parameter. Already some such enzymes are being investigated for this purpose. *Sulfolobus solfataricus* can be grown at 87°C. Its alcohol dehydrogenase is active in the range 30-95°C and activity increases continuously with temperature. Buffer, pH and protein concentration greatly influence the alcohol dehydrogenase's thermal stability. Stability decreased with increasing temperature: the enzyme was stable for several days at 50°C but only for 3 hours at 85°C. (This is somewhat surprising, as the organism was grown at 87°C. This apparent poor stability at 85°C may indicate the presence of low molecular weight stabilizers in the intact bacterial cell.) In studies at 65°C, the alcohol dehydrogenase displayed some activity in the presence of acetonitrile, methanol, ethanol, propanol, ethyl acetate and dimethyl sulfoxide but scarcely any activity with dimethyl (or diethyl) formamide. The enzyme resisted proteolysis by trypsin, chymotrypsin and thermolysin.[69] Thioredoxins were purified from *S. solfataricus* and also from *Pyrococcus furiosus* (grown at 98°C) and *Bacillus acidocaldarius* (grown at 60°C). All three thioredoxins had good stability characteristics and significant activity at low temperature, indicating that they may have applications in bioelectronics.[69]

Bacteriorhodopsin is an integral membrane protein which acts in vivo as a light-driven proton pump in some photosynthetic bacteria. In vitro, it has a variety of potential technological applications. These include molecular electronics and nanometre-scale fabrications. Shen et al noted that bacteriorhodopsin could retain its native folded structure up to 140°C when incorporated into an ordered and multilayered self-assembled dry film. The 'dry' part is crucial: when hydrated, the two-dimensional lattice occurring in these sorts of film shows a (reversible) solid-liquid transition around 69°C and bacteriorhodopsin itself denatures irreversibly around 90°C. However, neither event occurs up to 140°C in dry films.[70] Given that bacteriorhodopsin is a membrane protein, it is perhaps quite reasonable that it should be stabilized by incorporation in a two-dimensional film. Nevertheless, the maintenance of the protein's structural integrity up to a temperature as high as 140°C is remarkable. Possibly, the dehydrated conditions prevailing within the dry film restrict molecular motions, so that neither the film lattice nor the bacteriorhodopsin itself can undergo the transitions which occur in the hydrated state.

'Manufactured' Thermozymes?

There may be scope to 'impose' enhanced thermophilicity on microorganisms. Bennett and collaborators investigated the evolutionary fitness of *E. coli* shifted from a growth temperature of 37°C to 42°C for up to 400 generations (roughly two months).[71] Absence of contamination by other strains or organisms was verified by the use of phenotypic markers. The so-called 'ancestral' line was well adapted to the culture medium used, increasing the chances of finding temperature-specific adaptations. From the evolutionary viewpoint, this gradual selection of spontaneous mutations is arguably superior to mutagenesis and screening, since any mutations having adverse pleiotropic effects will be avoided.

Single colony isolates were picked after 100, 200 and 400 binary fissions from each of the 42°C lines. These were placed in competition with the ancestral line (which carries a distinctive phenotypic marker) at 42°C. (Each line had been allowed a 24 hour preincubation at 42°C to ensure that acclimatization had taken place before competition was imposed.) Fitness was defined as the ratio of the number of doublings for the 42°C and ancestral clones. No significant differences in fitness occurred after 100 generations. After 200 and 400 binary fissions, however, a 7-8% increase in fitness was observed relative to the ancestral clone. The 42°C lines also showed extra fitness at 37°C in competition with the ancestor.[71]

One must note that this study observed the viability of a whole organism, rather than the stability of a particular protein. The increases in fitness could have arisen from a number of possible adaptations which may or may not be related to a critical protein. Also, the temperature shift employed was mild, remaining within the mesophilic range. Nevertheless, such a strategy could have possibilities for the isolation of a 'thermophilized' protein where strain genotype and selection regime are well-defined.

Conclusion

Thermophilic proteins appear to have less conformational flexibility and less catalytic efficiency, at least at mesophilic temperatures, than do their mesophilic counterparts.[38] The extra stability conferred by the greater molecular rigidity of thermophilic proteins extracts a price since the thermophilic enzyme is less able (at mesophilic temperatures) to undergo the conformational flexions required for substrate torsion during catalysis.

A major difference between mesophilic and thermophilic proteins is that the latter show less tendency to unfold. The extra conformational stability arises from formation of a small number of additional salt bridges, hydrogen bonds, aromatic interactions or hydrophobic interactions. It seems that thermophilicity arises mainly from the prevention of unfolding and the ability of the native polypeptide conformation to shield 'thermolabile' residues (chapter 2) from inactivation. Once unfolded, the polypeptide will immediately undergo deleterious covalent reactions, resulting in irreversible inactivation. Jaenicke has reviewed molecular aspects of protein adaptation to extreme conditions and refers to the difficulties in ascertaining what changes lead to halo- or barophilicity or to toleration of extremes of pH[68] (echoing ref. 37). Further industrial and process applications of thermophilic enzymes are described in refs. 72 and 73. These applications may be further extended in the future, as Hei and Clark have reported that some thermophilic proteins (at least) are stabilized by high pressures; this effect may be due to hydrophobic interactions.[74] (There is a brief discussion of pressure effects on mesophilic protein stability in chapter 10.)

References

1. Brock TD. Thermophilic Organisms and Life at High Temperatures. Berlin: Springer Verlag, 1978.
2. Brock TD, ed. Thermophiles: General, Molecular and Applied Microbiology. NY: John Wiley and Sons, 1986.
3. Robb FT, Place AR, eds. Archaea: A Laboratory Manual. Volume 1, Thermophiles. Cold Spring Harbor Laboratory Press, 1996.
4. Sowers KR, Schreier HJ, eds. Archaea: A Laboratory Manual. Volume 2, Methanogens. Cold Spring Harbor Laboratory Press, 1996.
5. Das Sarma S, Fleischmann EM, eds. Archaea: A Laboratory Manual. Volume 3, Halophiles. Cold Spring Harbor Laboratory Press, 1996.
6. Brock TD. Thermophiles: General, Molecular and Applied Microbiology. NY: John Wiley and Sons, 1986.
7. Brock TD. Life at high temperatures. Science 1985; 230:132-138.
8. Kristjansson JK. Thermophilic organisms as sources of thermostable enzymes. Trends Biotechnol 1989; 7:349-355.
9. Windberger E, Huber R, Trincone A et al. *Thermotoga thermarum* sp. nov. and *Thermotoga neapolitana* occurring in African continental solfataric springs. Arch Microbiol 1989; 151:506-512.
10. Stetter KO, Fiala G, Huber G et al. Hyperthermophilic microorganisms. FEMS Microbiol Rev 1990; 75:2-3, 117-124.
11. Stetter KO, Huber R, Bloechl E et al. Hyperthermophilic archaea are thriving in deep North Sea and Alaskan oil reservoirs. Nature 1993; 365:743-745.
12. Kates M, Kushner D, Matheson AT, eds. The Biochemistry of the Archaea. (New Comprehensive Biochemistry 26.) Amsterdam: Elsevier, 1993.
13. Danson MJ, Hough DW, Lunt GG, eds. The Archaebacteria: Biochemistry and Biotechnology (Biochem Soc Symp 58). London: Portland Press, 1992.
14. Brock TD, Freeze H. *Thermus aquaticus* gen. n. and sp. n., a non-sporulating extreme thermophile. J Bacteriol 1969; 98:289-297.
15. Chien A, Edgar DB, Trela JM. Deoxyribonucleic acid polymerase from the extreme thermophile *Thermus aquaticus*. J Bacteriol 1976; 127:1550-1557.
16. Gelfand DH, Stoffel S, Lawyer FC et al. Purified thermostable enzyme. US Patent no. 4,889,818: 1989.
17. Hagen-Mann K. In: El-Gewely MR, ed. Site-Directed Mutagenesis and Protein Engineering. Amsterdam: Elsevier, 1991: 123-130.
18. Chaplin MF, Bucke C. Enzyme Technology. Cambridge UK: Cambridge University Press, 1990.
19. Imanaka T. Host-vector systems in thermophilic *Bacilli* and their applications. Trends Biotechnol 1983; 1:139-144.

20. Ward OP, Moo-Young M. Thermostable enzymes. Biotechnol Adv 1988; 6:39-69.
21. Vieille C, Zeikus JG. Thermozymes: identifying molecular determinants of protein structural and functional stability. Trends Biotechnol 1996; 14:183-190.
22. Russell RJM, Taylor GL. Engineering thermostability: lessons from thermophilic proteins. Curr Opin Biotechnol 1995; 6:370-374.
23. Voordouw G, Milo C, Roche RS. Role of bound calcium ions in thermostable proteolytic enzymes. Separation of intrinsic and calcium ion contributions to the kinetic thermal stability. Biochemistry 1976; 15:3716-3724.
24. Hensel R, Jakob I, Scheer H et al. Proteins from hyperthermophilic archaea: stability towards covalent modification of the peptide chain. In: Danson MJ, Hough DW, Lunt GG, eds. The Archaebacteria: Biochemistry and Biotechnology (Biochem Soc Symp 58). London: Portland Press, 1992:127-133.
25. Volkin DB, Middaugh CR. The effect of temperature on protein structure. In: Ahern TJ, Manning MC, eds. Stability of Protein Pharmaceuticals, Part A. Chemical and Physical Pathways of Protein Degradation. NY: Plenum Press Inc., 1992:215-247.
26. Perutz MF. Electrostatic effects in proteins. Science 1978; 201:1187-1191.
27. Russell RJM, Hough DW, Danson MJ et al. The crystal structure of citrate synthase from the thermophilic archaeon, *Thermoplasma acidophilum*. Structure 1994; 2:1157-1167.
28. Perutz MF, Raidt H. Stereochemical basis of heat stability in bacterial ferredoxins and in haemoglobin A2. Nature 1975; 255:256-259.
29. Zuber H. Temperature adaptation of lactate dehydrogenase: structural, functional and genetic aspects. Biophys Chem 1988; 29:171-179.
30. Suzuki Y. A general principle of increasing protein stability. Proc Japan Acad Sci 1989; 65 ser B:146-148.
31. Matthews BW, Nicholson H, Becktel WJ. Enhanced protein stability from site-directed mutations that decrease the entropy of unfolding. Proc Nat Acad Sci USA 1987; 84:6663-6667.
32. Creighton TE. Proteins: Structures and Molecular Properties, 2nd edn. NY: WH Freeman, 1993:181, 311-312.
33. Watanabe K, Masuda T, Ohashi H et al. Multiple proline substitutions cumulatively thermostabilize *Bacillus cereus* ATCC7064 oligo-1,6-glucosidase. Eur J Biochem 1994; 226:277-283.
34. Kizaki H, Hata Y, Watanabe K et al. Polypeptide folding of *Bacillus cereus* ATCC7064 oligo-1,6-glucosidase revealed by 3.0Å resolution X-ray analysis. J Biochem (Tokyo) 1993; 113:646-649.
35. Watanabe K, Chishiro K, Kitamura K et al. Proline residues responsible for thermostability occur with high frequency in the loop regions of an extremely thermostable oligo-1,6-glucosidase from *Bacillus thermoglucosidasius*. J Biol Chem 1991; 266:24287-24294.
36. Hensel R. Proteins of Extreme Thermophiles. In: Kates M, Kushner DJ, Matheson AT, eds. The Biochemistry of Archaea (Archaebacteria). Amsterdam: Elsevier, 1993:209-221.
37. Daniel RM, Dines M, Petach HH. The denaturation and degradation of stable enzymes at high temperatures. Biochem J 1996; 317:1-11.
38. Fontana A. Structure and stability of thermophilic enzymes. Studies on thermolysin. Biophys Chem 1988; 29: 181-193.
39. Mozhaev VV, Martinek K. Structure-stability relationships in proteins: new approaches to stabilizing enzymes. Enzyme Microb Technol 1984; 6:50-59.
40. Varley PG, Pain RH. Relation between stability, dynamics, and enzyme activity in 3-phosphoglycerate kinases from yeast and *Thermus thermophilus*. J Mol Biol 1991; 220:531-538.

41. Wrba A, Schweiger A, Schultes V et al. Extremely thermostable D-glyceraldehyde-3-phosphate dehydrogenase from the eubacterium *Thermotoga maritima*. Biochemistry 1990; 29:7592-7594.
42. O'Brien AM, Ó Fágáin C. Chemical stabilization of recombinant horseradish peroxidase. Biotechnol Techniques 1996; 10:905-910.
43. Chan MK, Mukund S, Kletzin A et al. Structure of a hyperthermophilic tungstopterin enzyme, aldehyde ferredoxin oxidoreductase. Science 1995; 267:1463-1469.
44. Eijsink VGH, Vriend G, Hardy F et al. Structural determinants of the thermostability of thermolysin-like *Bacillus* neutral proteases. In: van den Tweel WJJ, Harder A, Buitelaar RM, eds. Stability and Stabilization of Enzymes. Amsterdam: Elsevier, 1993:91-99.
45. Vihinen M, Mäntsälä P. Microbial amylolytic enzymes. CRC Crit Rev Biochem Mol Biol 1989; 24:329-418.
46. Vihinen M, Olikka P, Niskanen J et al. Site directed mutagenesis of a thermostable alpha-amylase from *Bacillus stearothermophilus*: putative role of three conserved residues. J Biochem (Tokyo) 1990; 107:267-272.
47. Yankov D, Dobreva E, Beschkov V et al. Study of optimum conditions and kinetics of starch hydrolysis by means of thermostable alpha-amylase. Enzyme Microb Technol 1986; 8:665-667.
48. Joyet P, Declerck N, Gaillardin C. Hyperthermostable variants of a highly thermostable alpha-amylase. Bio/Technology 1992; 10:1579-1583.
49. Declerck N, Joyet P, Gaillardin C, Masson J-M. Use of amber suppressors to investigate the thermostability of *Bacillus licheniformis* alpha-amylase. J Biol Chem 1990; 265:15481-15488.
50. Horinouchi S, Fukusumi S, Ohshima T et al. Cloning and expression in *E. coli* of two additional amylase genes of a strictly anaerobic thermophile, *Dictyoglomus thermophilum*, and their nucleotide sequences with extremely low guanine-plus-cytosine contents. Eur J Biochem 1988; 176:243-253.
51. Saha BC, Zeikus JG. Novel highly thermostable pullulanase from thermophiles. Trends Biotechnol 1989; 7:234-239.
52. Saha BC, Mathupala SP, Zeikus JG. Purification and characterization of a highly thermostable novel pullulanase from *Clostridium thermohydrosulfuricum*. Biochem J 1988; 252:343-348.
53. Meng M, Bagdasarian M, Zeikus JG. The role of active-site aromatic and polar residues in catlysis and substrate discrimination by xylose isomerase. Proc Nat Acad Sci USA 1993; 90:8459-8463.
54. Dekker K, Yamagata H, Sakaguchi K et al. Xylose (glucose) isomerase gene from the thermophile *Thermus thermophilus*: cloning, sequencing and comparison with other thermostable glucose isomerases. J Bacteriol 1991; 173:3078-3083.
55. Inyang CU, Gebhart U, Obi SKC et al. Isolation and characterization of a D-glucose/xylose isomerase from a new thermophilic strain *Streptomyces* sp. (PLC). Appl Microbiol Biotechnol 1995; 43:632-638.
56. Vieille C, Hess JM, Kelly RM et al. *xylA* Cloning and sequencing and biochemical characterization of xylose isomerase from *Thermotoga neapolitana*. Appl Envir Microbiol 1995; 61:1867-1875.
57. Meng M, Bagdasarian M, Zeikus JG. Thermal stabilization of xylose isomerase from *Thermoanaerobacterium thermosulfurigenes*. Bio/Technology 1993; 11: 1157-1161.
58. Merkler DJ, Srikumar K, Marchese-Ragona SP et al. Aggregation and thermoinactivation of glutamine synthetase from an extreme thermophile, *Bacillus caldolyticus*. Biochim Biophys Acta 1988; 952:101-114.

59. Marino G, Nitti G, Arnone MI et al. Purification and characterization of aspartate aminotransferase from the thermophilic archaebacterium *Sulpholobus solphataricus*. J Biol Chem 1988; 263: 12305-12309.
60. Fabry S, Hensel R. Purification and characterization of D-glyceraldehyde-3-phosphate dehydrogenase from the thermophilic archaebacterium *Methanothermus fervidus*. Eur J Biochem 1987; 165:147-155.
61. Ohshima T, Soda K. Thermostable amino acid dehydrogenases: applications and gene cloning. Trends Biotechnol 1989; 7:210-214.
62. Nagata S, Tanizawa K, Esaki, N et al. Gene cloning and sequence determination of leucine dehydrogenase from *Bacillus stearothermophilus* and structural comparison with other NAD(P)$^+$-dependent dehydrogenases. Biochemistry 1988; 27:9055-9062.
63. Ohshima T, Wandrey C, Kula M-R et al. Improvement for L-leucine production in a continuously operated enzyme membrane reactor. Biotechnol Bioeng 1985; 27:1616-1618.
64. Wrba A, Jaenicke R, Huber R et al. Lactate dehydrogenase from the extreme thermophile *Thermotoga maritima*. Eur J Biochem 1990; 188:195-201.
65. Christakopoulos P, Macris BJ, Kekos D. Exceptionally thermostable alpha- and beta-galactosidases from *Aspergillus niger* separated in one step. Process Biochem 1990; 25:210-212.
66. Pisani FM, Rella R, Raia CA et al. Thermostable beta-galactosidase from the archaebacterium *Sulpholobus solphataricus*. Eur J Biochem 1990; 187: 321-328.
67. Tayeh MA, Madigan MT. Malate dehydrogenases in phototrophic purple bacteria. Biochem J 1988; 252:595-600.
68. Jaenicke R. Protein stability and molecular adaptation to extreme conditions. Eur J Biochem 1991; 202:715-728.
69. Raia CA, D'Auria S, Guagliardi A et al. Characterization of redox proteins from extreme thermophilic archaebacteria: studies on alcohol dehydrogenase and thioredoxins from *Pyrococcus furiosus, Sulfolobus solfataricus* and *Bacillus acidocaldarius* and potential role in bioelectronics. Biosensors Bioelectron 1995; 10:135-140.
70. Shen Y, Safinya CR, Liang KS et al. Stabilization of the membrane protein bacteriorhodopsin to 140°C in two-dimensional films. Nature 1993; 366:48-50.
71. Bennett A, Dao K, Lenski R. Rapid evolution in response to high-temperature selection. Nature 1990; 346:79-81.
72. Coolbear T, Daniel RM, Morgan HW. The enzymes from extreme thermophiles: bacterial sources, thermostabilities and industrial relevance. In: Fiechter A, ed. Advances in Biochemical Engineering and Biotechnology 45. Berlin: Springer Verlag, 1992:57-96.
73. Adams MWW, Kelly RM, eds. Biocatalysis at Extreme Temperatures (ACS Symposium Series 498) Washington DC: American Chemical Society, 1992.
74. Hei DJ, Clark DS. Pressure stabilization of proteins from extreme thermophiles. Appl Envir Microbiol 1994; 60:932-939.

Part 2:
Manipulating Protein Stability

CHAPTER 5

Use of Stabilizing Additives

Introduction

It has long been known that inclusion of low molecular weight substances such as glycerol or sucrose in protein solutions can greatly stabilize the critical protein's biological activity. A variety of compounds (often used at concentrations of 1M or greater) can increase the stability of proteins in solution or those undergoing processes such as freeze drying. The range of such compounds is very wide. It includes substances as diverse as substrates (and specific ligands), salts, glycerol, sucrose, polyethylene glycol, chelating agents, reducing agents and proteins such as serum albumin.[1-4]

However, it took some time for the exact mechanism of this stabilization to be ascertained. Timasheff and colleagues have shown that these types of substances are preferentially excluded from the vicinity of the protein molecules, since their binding to the protein is thermodynamically unfavorable. The protein molecule is preferentially hydrated by the solvent water. Loss of the protein's compact, properly folded structure (denaturation) will increase the protein-solvent interface. This in turn will tend to increase the degree of thermodynamically unfavorable interaction between the additive and the protein molecule. The result is that the protein molecule is stabilized by the additive.[5,6] This 'preferential exclusion' means that there is less of the solute (additive) immediately surrounding the protein than there is in the bulk solution; it does not necessarily mean that *no* solute molecules can penetrate the protein molecule's hydration shell.[7] Timasheff discusses many aspects of solute/cosolvent interactions with proteins in ref. 5 and gives both a theoretical framework for their effects and some practical hints.

Types of Stabilizing Molecules

A particular salt exerts stabilizing or destabilizing effects on proteins depending on its position in the Hofmeister lyotropic series which relates to the effects of salts on protein solubility:[1,4]

$(CH_3)_4N^+ > N^+H_4 > K^+, Na^+ > Mg^{2+} > Ca^{2+} > Ba^{2+}$

$SO_4^{2-} > Cl^- > Br^- > NO_3^- > ClO_4^- > SCN^-$

Here both cations and anions are ranked in order of their stabilizing effects: the most stabilizing ions lie on the left-hand side while those towards the right become progressively more destabilizing. The stabilizing ions 'salt out' hydrophobic residues in the protein, causing the adoption of a more compact structure. This effect may be attributed to the increased ionic strength of the solution and to the increase in the number of water clusters around the protein.[4] This helps prevent

Stabilizing Protein Function, by Ciarán Ó Fágáin.
© Springer - Verlag and Landes Bioscience 1997.

the unfolding which is the initial event in any protein deterioration process (see chapters 1 and 2).

Most stabilizing ions seem to act via a surface tension effect.[5,6] Ions can also stabilize proteins by shielding surface charges and can act as osmolytes (see 'Empirical Considerations' below) by affecting the bulk properties of water.[8] Note that ammonium sulfate, which is widely used as a stabilizing additive and as a non-inactivating precipitant, comprises two of the most stabilizing ions from the Hofmeister list above, the NH_4^+ cation and the SO_4^- anion. One can stabilize proteins in solution while avoiding precipitation by adding ammonium sulfate to a final concentration in the range 20-400 mM.[8] Besides ammonium sulfate, salts containing citrate, sulphate, acetate, phosphate and quaternary ammonium ions are generally useful.[8] Note, however, that the nature of the counterion will influence the overall effect of such compounds on protein stability.[5,6] (This assumes that ions used or added do not act as substrates, activators or inhibitors of the enzyme or protein in question and that added ions do not interfere with, or precipitate, possibly-essential ions already in solution. Ammonium ion, for instance, is actually a substrate for the enzyme glutamate dehydrogenase.)

Glycerol, sugars and polyethylene glycol are polyhydroxy compounds. They can form many hydrogen bonds and aid formation of a 'solvent shell' around the protein molecule that is distinct from the bulk aqueous phase. They can also increase the surface tension and viscosity of a solution.[4]

Chelating agents act to complex metal ions. By doing this, they prevent oxidation by active oxygen species.[1] They can also prevent metal ion-induced aggregation. They can, however, remove a catalytically-essential metal ion from an active site, leading to loss of activity.

Reducing agents prevent the destructive oxidation of essential structural or functional features (and chelating agents can prevent metal ion-mediated oxidations),[9] but they do have potential drawbacks. The commonly used 2-mercaptoethanol can reduce disulfide bonds in proteins and also catalyzes a thiol-disulfide exchange[3] which may lead to aggregation. Tombs has highlighted this particular property and cautions against the widespread use of 2-mercaptoethanol in cases where no direct benefits have been proven.[10]

Basis of Stabilization: Empirical Considerations

Volkin and Klibanov have classified stabilizing additives into three classes: specific (substrates and ligands, where the *Native* $\rightarrow$ *Unfolded* equilibrium shifts towards the native form), nonspecific (neutral salts and polyhydroxyl compounds, which function as explained above) and competitors (which 'out compete' the enzyme for the inactivating reagent or remove the catalysts of deteriorative chemical reactions: examples include added protein, chelating and reducing agents).[4] Schein has reviewed solvents' effects on proteins, including aggregation and precipitation. She stresses the importance of the hydration or solvent shell surrounding protein molecules in solution[8] and has divided solutes into osmolytic and ionic stabilizers. Osmolytes affect solvent viscosity and surface tension: thus, they influence solvent ordering (See also ref. 6). They are not highly charged and do not affect enzyme activity up to 1 M concentration. Osmolytes include polyols, sugars, polysaccharides, neutral polymers and amino acids. Glycerol is the most widely used osmolyte. It appears actually to enter into the lattice structure of water. In this way it strengthens the solvent lattice structure surrounding the protein molecule and thus stabilizes the protein. The principal method of protein stabilization by ionic compounds is by shielding of surface charges. Schein also gives a useful summary of the mode of action and suitable concentration ranges of osmolytes, ions, reductants, protease inhibitors and bacteriocides.[8]

Carpenter and Crowe have undertaken a detailed investigation of solute protection of lactate dehydrogenase during freeze-thawing, using a total of 28 solutes from widely different chemical classes, including sugars, polyols, amino acids, methylamines and lyotropic salts.[7] All except NaCl had some stabilizing effect on the enzyme. The only property shared by all of these compounds is that all have been shown to be preferentially excluded from contact with the protein surface in aqueous solution. Solutes such as urea, which destabilize proteins in solution by binding preferentially to the protein surface, were also found to destabilize lactate dehydrogenase during freeze-thawing. These results accord with those noted by Timasheff and Arakawa.[6]

Heavy water (D_2O or ^2H_2O) can stabilize both enzymes and antibodies. The effect is seen at quite low concentrations and is not due to the lesser dissociation of heavy water, since stabilization occurred at both pH 7.5 and at pH 8.2.[11] Heavy water had a similar stabilizing effect in the case of butyrylcholinesterase thermoinactivation at pH 7.[12]

Combes and colleagues have used Raman spectroscopy as a non-invasive technique to study additive-water and additive-enzyme interactions. One can gain information on both intermolecular bonds (between solvent molecules) and on intramolecular bonds (within a protein conformation) by Raman methods.[13] These researchers determined the Raman spectra of aqueous solutions of alkali metal halides (i.e. the fluorine, chloride and bromide salts of lithium, sodium and potassium). Anions seemed to exert a greater influence than cations on the structural organization of water, and larger ions were more disruptive of this structure. These changes in the organization of water molecules likely explain the non-specific effects of salts on enzyme stability. Stabilizing polyhydric alcohols (erythritol, ethylene glycol, glycerol, sorbitol and xylitol) had much less effect on water's structural organization, suggesting that their protective effects arise from direct interactions (whether specific or non-specific) with enzyme polypeptides.

Combes and colleagues took this work further by determining the Raman spectra of two enzymes, invertase and lysozyme, in both deuterium oxide (heavy water) and water. This allowed assignment of particular spectral peaks to the various functional groups. Amide I and amide III are two distinct band regions within a polypeptide Raman spectrum. Their respective bandwidths, and especially the distance between them, correlate with the different types of secondary structure (alpha helix, beta sheet, beta turn and random coil). A decreased amide I-amide III distance implies a more ordered, alpha-helical structure and, therefore, a stronger overall tertiary conformation. A greater amide I-amide III distance, on the other hand, suggests a less-ordered, random coil-type secondary structure and a 'softer' tertiary conformation.[13] Therefore, changes in the Raman spectrum's amide I-amide III distance (wavenumbers, cm^{-1}) in the presence of additives reflect changes in protein structure. For lysozyme, the concentration-dependent protecting effects of alkali halide salts correlated with the amide I-amide III distance: increased thermal stability was matched by enhanced conformational stability. (KBr at a concentration of 0.5 M was the most stabilizing of the reported combinations.) NaCl protection of invertase yielded a bell-shaped curve. The degree of protection was much more pronounced than that of KBr on lysozyme and maximal protection occurred at 2.0 M NaCl, corresponding with the minimal amide I-amide III distance.

NaCl concentrations of 1.0 and 4.0 M had identical effects on thermal stability and amide I-amide III distance. The effects of KBr on invertase were also investigated by Raman spectroscopy. (KBr is a much less effective protector of invertase activity than NaCl, despite KBr's greater effect on water's structural organization.) Once again the concentration-dependent minimal amide I-amide III distance corresponded with the maximal protection effect. However, there was no quantitative relationship between the amide I-amide III distance and the protecting effect of *different* salts. NaCl at 2.0 M and KBr at 0.1 M gave virtually identical amide I-amide III values for invertase but 2.0 M NaCl's protecting effect was nearly 50 times greater. Finally, the effects of three potassium halides on invertase were compared at their respective optimally-stabilizing concentrations. KCl (3.0 M) was by far the most effective, perhaps due to a specific effect on invertase association, while KBr (0.5 M) was more protecting than KF (2.0 M). There was a qualitative relationship between stabilizing effect and the amide I-amide III distance.[13]

A Case Study: Lysozyme

Chapter 2 outlined the influential studies of Ahern and Klibanov concerning the molecular events leading to inactivation of lysozyme. Tomizawa and colleagues have revisited this subject, focusing on the inactivation processes at neutral pH over a wide range of protein concentrations and particularly on the ability of certain additives to prevent or limit the deleterious chemical reactions leading to lysozyme inactivation.[14,15] These workers established that the inactivation of hen egg white lysozyme at pH 4 and 100°C was not due to a single chemical reaction. As a number of deleterious chemical reactions accumulated in the lysozyme molecule, the Gibbs energy difference between folded and unfolded forms decreased markedly. This occurred to such an extent that some molecules were in an unfolded state even at low temperatures and, accordingly, were without activity.[16] At pH 6, activity loss at 100°C did not have first-order characteristics. There was a dependence on protein concentration and precipitates were noted even in the early stages of inactivation experiments at high protein concentrations. These observations indicated that intermolecular events contribute significantly to the inactivation process at pH 6.[14] While deamidation of Asn residues and rupture of Asp peptide bonds on their C-terminal side (chapter 2) are intramolecular reactions within a single polypeptide, disulfide exchange reactions can be intermolecular and may lead to aggregation. Tomizawa and colleagues inferred that disulfide exchange, in addition to the other inactivating chemical reactions, could occur in lysozyme at pH values as low as 6. They tested this possibility by including $CuCl_2$ in their inactivation mixtures (in 2.5-fold molar excess over the lysozyme concentration, which was held at $1 mg.ml^{-1}$). $CuCl_2$ is known to inhibit disulfide exchange[17] and its inclusion had a notable effect on the inactivation time course of lysozyme. In the presence of $CuCl_2$, inactivation rates were considerably lessened, did not depend on protein concentration and followed first-order kinetics. Thus the inclusion of copper ions effectively prevented the disulfide-mediated polymolecular reactions which led to aggregation and precipitation.

Inclusion of 8 M acetamide (roughly 50% v/v) was also beneficial at pH 6 and 100°C. This reagent suppresses the Asn deamidation reaction but promoted the formation of succinimide derivatives of lysozyme[14] which possess activity levels above those of native lysozyme.[18] In model experiments, acetamide was shown to raise the pK_a values of both the carboxyl function and of the Cys thiol group by at least one pH unit. The effect of this is to shift the inactivation mechanism to that which would occur under more acidic conditions.[14] While inactivation in 8 M acetamide was always first-order (and thus monomolecular) inclusion of copper ion had an additional protecting effect, indicating that *intra*molecular disulfide exchange also occurred. Ethanol was also valuable as a protecting additive: inactivation at pH 6 and 100°C in 50% ethanol was slower than in purely aqueous buffer and the inactivation characteristics resembled those in 8 M acetamide. However, copper ion was ineffective in 50% ethanol. Ethanol is thought to exert its protecting effect through suppression of intermolecular interactions and alteration of the pK_a values of dissociable residues, as does acetamide. The extent of protection against inactivation correlated with the degree of pK_a shift induced by the particular additive: 50% glycerol was a much less effective protectant of lysozyme than acetamide or ethanol; it also had much less effect on the pK_a values of the important functional groups.[14] Acetamide and copper ions, when included in lysozyme solutions inactivated at 100°C and pH 8, had similar protective effects to those observed at pH 6.[14]

Tomizawa and colleagues also focused on Asp/Asn racemization in lysozyme at pH 6 and 100°C and its prevention by additives. These deleterious reactions were promoted by phosphate ions: deamidation rates for a Asn-Gly pair in a lysozyme-derived peptide were 7-fold less in MOPSO than in phosphate buffer of the same pH and concentration.[15] Various salts (especially LiCl) opposed this effect of phosphate ions but had no protecting effect themselves in the absence of phosphate. Thus the salts' protection of lysozyme activity was due to their antagonizing the phosphate-catalyzed racemization. Trifluoroethanol (TFE, 5%) also helped stabilize lysozyme, probably by suppressing the deamidation of Asn residues in the lysozyme polypeptide. TFE was somewhat less effective in suppressing the racemization reaction, merely opposing the effect of phosphate ions. This finding implies that racemization can result also from the succinimide intermediate which occurs in the deamidation reaction. (TFE at 5% concentration has been shown to suppress deamidation of an Asn-Gly pair in a lysozyme-derived oligopeptide; its effectiveness in this regard is several hundred-fold greater than the antagonism by salts of the phosphate-catalyzed Asn-Gly deamidation. This implies that the protecting effect of TFE goes beyond simple antagonism of the phosphate catalysis and that it also suppresses a nucleophilic attack by the backbone N-atom of the residue on the C-terminal side of the vulnerable Asn [Gly in the case of the Asn-Gly pair]. Also, TFE exerted a small protecting effect even at 1% concentration.[15]) Phosphate ions promoted inactivation of lysozyme at pH 6 and 100°C but not in a concentration-dependent manner. Half lives were much longer in MOPSO (0.05M, equivalent to the phosphate concentration) but could be increased a little further by inclusion of 0.5 M LiCl. This differed from results for deamidation of the Asn-Gly pair and suggested that more than one chemical reaction is responsible for the inactivation of lysozyme (although inactivation correlated more closely with the extent of deamidation [$r = 0.914$] than of racemization [$r = 0.690$]). Differential scanning calorimetry was used to measure lysozyme thermal stabilities in the presence of these various additives. The substances giving greatest protection against thermoinactivation —TFE and LiCl—actually led to slight *decreases* in T_m, indicating that their stabilizing effects on lysozyme arose from suppression of deleterious chemical reactions and not to increased conformational stability. These findings have implications for the widespread use of phosphate salts as buffers for protein purification and assay. Researchers may unwittingly expose their target protein to the possibility of adverse chemical reactions catalyzed by phosphate ions.[15] While the studies of Tomizawa et al focused on the high-temperature effects of phosphate, one should note that sodium phosphate salts may also be contraindicated in low temperature operations such as freeze drying (see chapter 9).

Gekko conducted a thermodynamic study of lysozyme stabilization by low-molecular-weight polyols. Denaturation temperature, T_d, increased with the polyol concentration and with the number of hydroxymethyl groups per polyol molecule. Meanwhile, denaturation enthalpy increased with the polyol concentration but did not vary with polyol chain length. Results agreed with the hypothesis that polyol stabilization of lysozyme is due to a strengthening of the protein's hydrophobic interactions by a preferential solvent interaction effect.[19]

Other Examples

Tams and Welinder have studied the effects of organic and ionic additives on the unfolding/refolding of *Coprinus cinereus* peroxidase, an enzyme containing both heme and calcium ions.[20] Its unfolding and refolding were measured at alkaline pH (12.1), in 4 M urea and at 61.2°C. Two distinct methods, thermolysin digestion and EDTA experiments (both at 59.4°C), were used to measure the unfolding rate constant, and gave identical values (1.5 ms^{-1}). Thermolysin can degrade the unfolded apoprotein to peptides but not the native holoprotein. EDTA chelates the calcium released from the peroxidase as it unfolds, thus effectively rendering the unfolding process irreversible. Accordingly, the unfolding rate may be measured independently of refolding. For refolding experiments, an excess of calcium was added. An isosbestic point in the peroxidase absorption spectra indicated that a two-state process was operative: heme was either bound in the native fashion or it was completely released from the polypeptide. (Multiphasic behavior was observed under some of the experimental conditions employed, however: in urea and at high pH, both unfolding and refolding were biphasic. Also, the reversible unfolding at 61.2°C did not fit a two-state process.) The effects of five additives (glycerol, ethanol, sodium sulfate, guanidinium chloride and sodium chloride, all at 100 mM concentration) on unfolding and refolding of *C. cinereus* peroxidase were monitored. At high pH, salts destabilized the enzyme, mainly by increasing the unfolding rate. The greater effects of salts on the unfolding process suggests that the pH 12.1 transition state resembles U more than N (in the standard $N \leftrightarrow U \rightarrow I$ model; chapter 1) and that the unfolded enzyme (U) at pH 12.1 is in a compact denatured state. Glycerol had the opposite effects and stabilized the enzyme at high pH. Unfolding at this high pH value likely involves the loss of protons from Lys and Arg residues and the ionization of tyrosines. In urea, guanidinium chloride increased the rate of unfolding but the other additives had little effect on the unfolding rate in urea. Glycerol and sodium sulfate increased the *re*folding rate in urea, unlike ethanol and guanidinium chloride, which decreased the rate of refolding. The greater effects of the additives on the refolding rates in urea indicate that the peroxidase transition state in urea more closely resembles N than U, unlike the situation at pH 12.1. At high temperature, only guanidinium chloride and sodium chloride, in that order, exerted any notable effect on peroxidase unfolding (both enhanced the unfolding rate). Glycerol and sodium sulfate enhanced the refolding rate at high temperature, in contrast to the other three additives which greatly decreased the refolding process. For *C. cinereus* peroxidase at high temperature, sodium chloride increases the rates of both unfolding and refolding. The unfolding and refolding processes of this enzyme are similar at high temperature and in urea but different at high pH. It seems that electrostatic interactions have an important effect on the stability of this type of heme-containing peroxidase. Some of these enzymes, including the present *C. cinereus* and horseradish peroxidases (chapter 6) have commercial potential or use and have been expressed in recombinant form. Tams and Welinder's work indicates a means of achieving greater yields of such recombinant peroxidases by the inclusion of glycerol in the folding cocktail and the use of a low ionic strength.[20]

NAD^+ has been shown to stabilize tetrameric lactate dehydrogenase in proportion to its occupation of the active sites of the tetramer. Formation of an abortive LDH-NAD^+-pyruvate complex exerted a stabilizing effect but only if all four active sites of the tetramer were simultaneously occupied.[21] Numerous other examples of substrate stabilization are known. Quinine sulphate has been reported to be particularly effective in protecting glucoamylase against hydrogen peroxide oxidation.[22] The use of a variety of colloids and high polymers to stabilize industrial enzymes has been summarized by Barker.[23]

Use of Additives

It is important to note that the additives discussed below are *generally* applicable as stabilizing agents for proteins but a given substance may not be effective for a particular protein. Both sucrose and PEG, for instance, are good stabilizers of invertase but have denaturing effects on lysozyme; the same additive has contrary effects on the two enzymes.[24] One should note the stabilizing or destabilizing effects of the component ions of a salt when choosing additives; ref. 6 includes a useful discussion of this topic.

Osmolytes are a diverse group of substances comprising such compounds as polyols, mono- and polysaccharides, neutral polymers (such as PEG) and amino acids and their derivatives. Their general properties have been outlined above and are described in ref. 9. One should use polyols and sugars at high final concentrations: typical figures range from 10-40%(w/v).[5,6,8] Sugars are reckoned to be the best stabilizers[5,6] but reducing sugars can react with protein amino groups, leading to inactivation.[25] This problem can be avoided by using non-reducing sugars or the corresponding sugar alcohols. Glycerol is a very widely used low molecular weight polyol. Its advantages include its ease of removal by dialysis and its non-interference with ion exchange chromatography.[8] However, glycerol suffers from two significant disadvantages: it is a good bacterial substrate[8] and it greatly lowers the glass transition temperature (T_g') of materials to be preserved by lyophilization or drying (see chapter 9). The 5-carbon sugar alcohol, xylitol, can often replace glycerol; it can be recycled from buffers and is not a convenient food source for bacteria.[8]

Polymers such as PEG are generally added to a final concentration of 1-15%(w/v). They increase the viscosity of the single phase solvent system and thereby help prevent aggregation. Note, however, that higher polymer concentrations will promote the development of a two phase system. The protein of interest will concentrate in one of these phases and this may actually lead to aggregation.[8]

Amino acids with no net charge, notably glycine and alanine, can act as stabilizers if used in the range 20-500 mM.[8] Amino acids and derivatives occur as osmolytes in nature.[6] Some related compounds, such as gamma-amino butyric acid (GABA) and trimethylamine N-oxide (TMAO), can be good stabilizers in the 20-500 mM range used for amino acids.[6,8]

Substrates and Specific Ligands

Addition of specific substrates, cofactors or competitive (reversible) inhibitors to purified proteins can often exert great stabilizing effects. (Indeed, their inclusion may be necessary where an essential metal ion or coenzyme is only loosely bound to the apoprotein.) Occupation of the target protein's binding or active site(s) by these substances leads to minor but significant conformational changes in the polypeptide backbone. The protein adopts a more tightly folded conformation, reducing any tendency to unfold[4] and (sometimes) rendering it less prone to proteolytic degradation. Occlusion of the protein's active site(s) by a bound substrate molecule or reversible inhibitor will protect those amino acid side chains which are critical for function. A starch degrading amyloglucosidase enzyme (from an *Aspergillus* species) stored in the presence of 14%(w/v) partial starch hydrolysate was 80% more stable over a 24-week period at ambient temperature than the corresponding enzyme preparation stored in the hydrolysate's absence.[26] Note that dialysis (or some other procedure for the removal of low molecular mass substances) may be necessary to avoid carry-over effects of the substrate or inhibitor when the protein is removed from storage for use in a particular situation where maximal activity is desired.

Use of Reducing Agents and Prevention of Oxidation Reactions

The thiol group of cysteine is prone to destructive oxidative reactions (chapter 2). One can prevent or minimize these by using reducing agents such as 2- (formerly beta-) mercaptoethanol (a liquid with an unpleasant smell) or dithiothreitol ('Cleland's reagent', or DTT, a solid with little odor). One should add 2-mercaptoethanol to reach a final concentration of 5-20 mM and then keep the solution under anaerobic conditions. To achieve these anaerobic conditions, one can gently bubble an inert gas such as nitrogen through the solution and fill it to the brim of a screwcap container to minimize headspace and the chances of gaseous exchange. Dithiothreitol is effective at lower concentrations: usually 0.5-1 mM will suffice.[3] Indeed, Schein has advised that the DTT concentration should not exceed 1 mM: it can act as a denaturant at higher temperatures and is not very soluble in high salt.[9] Note that these reducing agents are themselves prone to oxidation. (This is why solutions containing them must be stored under 'anaerobic conditions'.) Dithiothreitol oxidizes to form an internal disulfide which is no longer effective but which will not interfere with protein molecules.[3] 2-Mercaptoethanol, on the other hand, participates in intermolecular reactions and can form disulfides with protein thiol groups.[3] Such thiol-disulphide exchanges are highly undesirable and may actually lead to inactivation or aggregation. It is probably best to add reducing agents only in situations where they are known (or can be demonstrated) to be beneficial.[11]

Much of the oxidation of thiol groups is mediated by divalent metal ions which can activate molecular oxygen. Complexation of free metal ions (where they are not themselves essential for activity) can prevent destructive oxidation of thiol groups. EDTA may be used to complex metal ions.

Some substances can also protect against activity or potency losses during drying or freeze drying operations (see chapter 9). Gibson et al have devised and patented a polymer-carbohydrate system allowing long-term stabilization of analytical enzymes such as alcohol oxidase, horseradish peroxidase and alkaline phosphatase.[27,28] The carbohydrate portion is the sugar alcohol lactitol while the cationic, soluble polymer is typically DEAE (diethylaminoethyl)-dextran. Enzymes were dried under vacuum at 30°C in the presence or absence of 5% m/v lactitol and 1% m/v DEAE-dextran. The dried enzymes were incubated at 37°C for long periods. Unstabilized alcohol oxidase retained only 26% initial activity after 7 days while stabilized fractions retained 100% activity for up to 2 months.

There are alternative means of stabilizing proteins in the dry state. Glass has many useful properties but is not in fact a solid; rather it is a highly viscous liquid which has undergone a 'glass transition' to yield a brittle, amorphous solid. Franks has used stabilizing carbohydrate-based formulations which undergo a similar glass transition. These protect proteins during drying operations and the resulting solid effectively preserves the protein in a fully-active form. Since the glass formulation is water-soluble, addition of water allows one to redissolve the protein. Franks has summarized the underlying ideas in ref. 29 and discussed them in detail in refs. 30 and 31.

Interestingly, it appears that some hyperthermophilic Archaea use low molecular weight stabilizers in vivo. *Methanothermus fervidus* cells contain 1M K^+ and 0.3 M cyclic 2,3-diphosphoglycerate. The latter increases stability of *M. fervidus* enzymes in vitro and the *M. fervidus* organism increases its internal concentrations of K^+ and cyclic 2,3-diphosphoglycerate in response to an increase in growth temperature.[32,33] (Chapter 4 provides a detailed discussion of thermophilic proteins.)

Additive and solute use is undoubtedly important in the stabilization of enzymes for commercial and industrial use. Genuinely inventive combinations of stabilizing compounds have been devised (refs. 28, 31). Also, Modrovich has patented stabilized enzyme and coenzyme formulations, many of which are based on the use of high concentrations (up to 55%) of water-miscible polyol organic solvents. pH buffering reagents and bacteriostatic compounds are likewise often used. Modrovich has developed a variety of laboratory diagnostic products using this strategy.[34]

Additives can greatly enhance stability without chemical or genetic manipulation of the target protein. Using empirical knowledge and the principles described, one may devise a formulation applicable to many different proteins rather than just one. This is especially useful in drug applications, where the need to satisfy regulatory requirements may work against a chemically- or genetically-modified derivative. Of course, one should note that operations such as dilution, dialysis, buffer exchange or rehydration of a dry preparation will remove the stabilizing influences: one has achieved stability only in a particular context or ambience. In contrast, engineered proteins such as subtilisins (chapter 8) show increased stability under many different conditions.

References

1. Klibanov AM. Stabilization of enzymes against thermal inactivation. Adv Appl Microbiol 1983; 29:1-25.
2. Klibanov AM. Approaches to enzyme stabilization. Biochem Soc Transact 1983; 11:19-20.
3. Scopes RK. Protein Purification: Principles and Practice. 3rd edn. NY: Springer Verlag, 1994:317-324.
4. Volkin DB, Klibanov AM. Minimizing protein inactivation. In: Creighton TE, ed. Protein Function: a Practical Approach. Oxford: IRL Press, 1989:1-24.
5. Timasheff SN, Arakawa T. Stabilization of protein structure by solvents. In: Creighton TE, ed. Protein Structure: a Practical Approach Oxford: IRL Press, 1989: 331-345.

6. Timasheff SN. Stabilization of protein structure by solvent additives. In: Ahern TJ, Manning MC, eds. Stability of Protein Pharmaceuticals, Part B: *In Vivo* Pathways of Protein Degradation and Strategies for Protein Stabilization. NY: Plenum Press, Inc., 1992:265-285.
7. Carpenter JF, Crowe JH. The mechanism of cryoprotection of proteins by solutes. Cryobiology 1988; 25:244-255.
8. Schein CH. Solubility as a function of protein structure and solvent components. Bio/Technology 1990; 8:308-317.
9. Klibanov AM, Kaplan NO, Kamen ND. Chelating-agents protect hydrogenase against oxygen inactivation. Biochim Biophys Acta 1979; 547:411-416.
10. Tombs MP. Stability of enzymes. J Appl Biochem 1985; 7:3-24.
11. Lemm U, Wenzel M. Stabilisierung von Enzymen und Antiseren durch schweres Wasser. Eur J Biochem 1981; 116:441-445.
12. Masson P, Laurentie M. Stability of butyrylcholinesterase: thermal inactivation in water and deuterium oxide. Biochim Biophys Acta 1988; 957:111-121.
13. Combes D, Auzanneau I, Zwick A. Thermal stability of enzymes: influence of solvatation medium (a Raman spectroscopy study). In: van den Tweel WJJ, Harder A, Buitelaar RM, eds. Stability and Stabilization of Enzymes. Amsterdam: Elsevier, 1993:29-36.
14. Tomizawa H, Yamada H, Tanigawa K et al. Effects of additives on irreversible inactivation of lysozyme at neutral pH and 100°C. J Biochem (Tokyo) 1995; 117: 369-373.
15. Tomizawa H, Yamada H, Wada K et al. Stabilization of lysozyme against irreversible inactivation by suppression of chemical reactions. J Biochem (Tokyo) 1995; 117:635-640.
16. Tomizawa H, Yamada H, Imoto T. The mechanism of irreversible inactivation of lysozyme at pH 4 and 100°C. Biochemistry 1994; 33:13032-13037.
17. Zale SE, Klibanov AM. Why does ribonuclease A irreversibly inactivate at high temperatures? Biochemistry 1986; 25: 5432-5444.
18. Tomizawa H, Yamada H, Ueda T et al. Isolation and characterization of 101-succinimide lysozyme that possesses the cyclic imide at Asp101-Gly102. Biochemistry 1994; 33:8770-8774.
19. Gekko K. Calorimetric study on thermal denaturation of lysozyme in polyol-water mixtures. J Biochem (Tokyo) 1982; 91:1197-1204.
20. Tams JW, Welinder KG. Unfolding and refolding of *Coprinus cinereus* peroxidase at high pH, in urea, and at high temperature. Effect of organic and ionic additives on these processes. Biochemistry 1996; 35:7573-7579.
21. Kube D, Shnyrov VL, Permyakov EA et al. Correlation between the thermal stability of the tetramer lactate dehydrogenase molecule from porcine muscles and the degree of occupation of its active sites by ligands. Biochemistry 1988; 52:962-970. (English translation of 'Biokhimiya', 1987)
22. Cho YK, Bailey JE. The influence of peroxide-stabilizing agents on enzyme deactivation by H_2O_2. Biotechnol Bioeng 1977; 19:157-158.
23. Barker SA. New approaches to enzyme stabilization. In: Wiseman A, ed. Topics in Enzyme and Fermentation Biotechnology. Chichester, UK: Ellis Horwood, 1982:68-78.
24. Combes D, Yoovidhya T, Girbal E et al. Mechanism of enzyme stabilization. Ann NY Acad Sci 1987; 501:59-62.
25. Quax WJ. Thermostable glucose isomerases. Trends Food Sci Technol 1993; 4:31-34.
26. Shah NK, Shah DN, Upadhyay CM et al. An economical, upgraded, stabilized and efficient preparation of amyloglucosidase. J Biotechnol 1989; 10:267-276.

27. Gibson TD, Higgins IJ, Woodward JR. Stabilization of analytical enzymes using a novel polymer-carbohydrate system and the production of a stabilized, single reagent for alcohol analysis. Analyst 1992; 117:1293-1297.
28. Gibson TD, Woodward JR. Enzyme stabilization. World Patent no. PCT/GB91/00443, Publ. No. W091/14773: 1991.
29. Franks F. Long-term stabilization of biologicals. Bio/Technology 1994; 12:253-256.
30. Franks F, Hatley RHM, Mathias SF. Materials science and the production of shelf-stable biologicals. Pharm Technol Int 1991; 3:24-34.
31. Franks F, Hatley RHM. Storage of materials. US Patent no. 5,098,893: 1992.
32. Hensel R. Proteins of Extreme Thermophiles. In: Kates M, Kushner DJ, Matheson AT, eds. The Biochemistry of Archaea (Archaebacteria). Amsterdam: Elsevier 1993:209-221.
33. Hensel R, Jakob I, Scheer H et al. Proteins from hyperthermophilic archaea: stability towards covalent modification of the peptide chain. In: Danson MJ, Hough DW, Lunt GG, eds. The Archaebacteria: Biochemistry and Biotechnology. (Biochemical Society Symposium vol 58). London: Portland Press Ltd, 1992:127-133.
34. Modrovich I. Enzymatic method and stabilized solutions for determining total cholesterol in human serum. US Patent no. 4,378,429: 1983.

CHAPTER 6

Chemical Modification of Proteins in Solution

Introduction: Scope of Chemical Modification

To describe the myriad possible chemical modifications of proteins would require a monograph devoted entirely to this topic, rather than just a single chapter in this book. This chapter attempts merely to survey the nature and types of chemical modifications available for the study and derivatization of enzymes and proteins. It describes the functional groups of proteins that are available for reaction and presents the types of reactions they undergo. It then summarizes a selection of chemical modifications which have stabilized a variety of enzymes and proteins.

Chemical modification procedures can provide important structural and functional information concerning a protein. Soluble enzymes (as distinct from immobilized ones, dealt with in chapter 7) can be chemically modified in a variety of ways so as to alter their properties. The chemistries of such reactions are well understood and a wide range of reagents has been available for many years (see, for example, refs. 1-4). Each of the 20 amino acids occurring in proteins has a free R-group or side chain. Many of these have reactive functional groups such as the thiol group of cysteine or the amino group of lysine residues. At least 9 amino acid side chains (Cys, Lys, Asp and Glu, Arg, His, Trp, Tyr and Met; ref. 1) can react with specific reagents under mild conditions to yield chemically-modified protein derivatives, often with altered properties. Inactivation is always a danger, since some amino acid side chains will be essential for function, but careful choice of modifying agent(s) and conditions can often minimize activity losses. The range of suitable reagents is constantly expanding due to the need for efficient conjugation or immobilization procedures. Table 6.1 lists reactive amino acid R-groups and some of the types of compounds useful for reactions with them. Useful reviews of the area include refs. 3 and 5-7. Ref. 5 includes outline protocols for the use of some of these reagents. The specific area of chemical stabilization of enzymes has been surveyed in refs. 5 and 7-10.

Chemical modification is in many ways complementary to site-directed mutagenesis and protein engineering (see chapter 8) as a methodology for the study of protein variants.[5,6] Relatively little structural information is required concerning the target protein. The experiments are often simple to carry out and protocols may be readily implemented. Thus, one can get results quickly. However, chemical modification of proteins is prone to a number of pitfalls (see ref. 3). These are also

Stabilizing Protein Function, by Ciarán Ó Fágáin.
© Springer - Verlag and Landes Bioscience 1997.

Table 6.1. ***Amino acid side chains and reagents for their modification***

Amino Acid	Side Chain	Reagents	Reaction	Ref
Cysteine	Thiol	5,5'-Dithiobis(2-nitrobenzoic acid)[1]	Disulphide formation	(a, b[1])
		Maleimido compounds	Alkylation	(c)
		p-Mercuribenzoate[1]	Addition	(b[1],c)
Lysine	Amino	Trinitrobenzene sulfonate[1]	Addition	(b[1], d[1])
		Imidates	Amidination	(e)
		Acid anhydrides	Acylation	(f)
		(Cyano)borohydride + aldehyde	Reductive alkylation	(g)
Arginine	Guanidino	Dicarbonyls	Not fully known	(b, h)
		9,10-Phenanthrenequinone[1]		(b[1])
		Camphorquinone-10-sulfonic acid		(i)
Histidine	Imidazole	Diethylpyrocarbonate[1]	Addition	(b[1],j)
Aspartic acid,	Carboxyl	Carbodiimides	Amidation	(k)
Glutamic acid	"	Trialkoxonium salts	Esterification	(l)
Tyrosine	Phenol	Tetranitromethane	Nitration	(m)
		Iodine	Iodination	(n)
Tryptophan	Indole	N-Bromosuccinimide	Oxidation	(o)
Methionine	Thioether	Hydrogen peroxide	Oxidation	(p)

[1] These compounds may be used for the spectrophotometric (or, in the case of Arg, fluorimetric) quantitation of the content of the reactive R-group in a protein: see respective references.

Note that reaction of Arg with the dicarbonyl 2,3-butanedione should be carried out in the dark (ref. b). This table presents only some of the more common reagents for protein R-group modification; it is by no means exhaustive.

(a) Riddles PW, Blakely RL, Zerner B. Reassessment of Ellman's reagent. Meths Enzymol 1983; 91:49-60.
(b) Roig MG, Kennedy JF. Perspectives for chemical modifications of enzymes. CRC Crit Rev Biotechnol 1992; 12:391-412.
(c) Riordan JF, Vallee BL. Reactions with *N*-ethylmaleimide and *p*-mercuribenzoate. Meths Enzymol 1972; 25:449-456.
(d) Fields R. The rapid determination of amino groups with TNBS. Meths Enzymol 1972; 25:464-468.
(e) Inman JK, Perham RN, DuBois GC et al. Amidination. Meths Enzymol 1983; 91:559-569.
(f) Klapper MH, Klotz IM. Acylation with dicarboxylic acid anhydrides. Meths Enzymol 1972; 25:531-536.
(g) Jentoft N, Dearborn DG. Protein labeling by reductive alkylation. Meths Enzymol 1983; 91:570-579.
(h) Yankeelov JA. Modification of arginine by diketones. Meths Enzymol 1972; 25:566-579.
(i) Pande CS, Pelzig M, Glass JD. Camphorquinoe-10-sulfonic acid and derivatives: convenient reagents for reversible modification of arginine residues. Proc Nat Acad Sci USA 1980; 77:895-899.
(j) Dominici P, Tancini B, Voltattorni CB. Chemical modification of pig kidney 3,4-dihydroxyphenylalanine decarboxylase with diethyl pyrocarbonate. J Biol Chem 1985; 260:10583-10589.
(k) Carraway KL, Koshland DE. Carbodiimide modification of proteins. Meths Enzymol 1972; 25:616-623.
(l) Wilcox PE. Esterification. Meths Enzymol 1972; 25:596-616.
(m) Riordan JF, Vallee BL. Nitration with tetranitromethane. Meths Enzymol 1972; 25:515-521.
(n) Morrison M. Iodination of tyrosine: isolation of lactoperoxidase (bovine). Meths Enzymol 1970; 17:653-660.
(o) Spande TF, Witkop B. Determination of the tryptophan content of protein with N-bromosuccinimide. Meths Enzymol 1967; 11:498-532.
(p) Neumann NP. Oxidation with hydrogen peroxide. Meths Enzymol 1972; 25:393-401.

mentioned in ref. 7 which stresses the importance of protecting active site residues by use of substrates or reversible inhibitors.

Two of the most reactive R-groups are the thiol and amino groups, which occur on the amino acids cysteine and lysine, respectively. These groups each react with a range of agents. Convenient quantitation of these two important groups can, in addition, be achieved chemically. Trinitrobenzenesulfonic acid (TNBS) will react with amino groups under alkaline conditions to yield a yellow compound which may be estimated spectrophotometrically at 420nm. Absorbance is proportional to the number of protein amino groups, and thus the number of free amino groups may be calculated with reference to a standard curve of A_{420} against the quantity of free amino group. Alpha-N-acetyl-L-lysine is a suitable compound for the preparation of a standard curve. Variants of this assay exist, but that described by Fields[11,12] is quite commonly used. Note that a free N-terminal amino group will also react with TNBS. Thiol groups may be spectrophotometrically estimated at 412nm following use of 5,5'-dithiobis(2-nitrobenzoic acid), also known as DTNB or Ellman's reagent, at neutral pH.[13] Note, however, that satisfactory access of the modifier to its target group cannot always be assumed: e.g. DTNB failed to detect any free -SH groups in native ovalbumin, but gave the correct number (four) when reacted with denatured ovalbumin.[2]

Active-site targeting represents a second application of chemical modification. R-group specific reagents may be used to identify residues in or near the active site. For instance, one can distinguish serine from cysteine proteinases by use of appropriate reagents. Many compounds have been designed to react specifically with the active site residues of particular enzymes. The chemically-modified residues may be conveniently identified by comparing the peptide map (or protein sequence) characteristics of the chemically-modified protein with those of the native. This approach can elucidate which active site residues are involved in binding and catalysis. Very many active-site directed reagents (or affinity labels) have been custom-made for particular enzymes. The kinetics of affinity label-induced inactivation can also provide information on the protein under investigation, as reviewed by Cardemil[14] and Rakitzis.[15] Eyzaguirre[16] and Shaw[17] contain useful descriptions of this specialized area of affinity labelling. These types of reagents are not used for purposes of protein stabilization and so will not be discussed further here.

Crosslinking of protein subunits or of adjacent proteins can be carried out with a wide variety of reagents. Crosslinking compounds of different length are available, whether homobifunctional or heterobifunctional (i.e. with similar or different reactive groups at either end of the molecule, respectively). Sometimes a group-specific function is combined with a highly-reactive, non-specific photoactivable group. Since the photoactivable group is tethered at one end of the crosslinker, it can react only with those protein R-groups within the length of the crosslinking molecule. Details and case summaries can be found in refs. 1-4 while crosslinkers in general are reviewed in ref. 18. Ever more sophisticated reagents are being synthesized and applied.[20-23] Many commercial catalogues have special sections devoted to such reagents, especially Pierce and Sigma.

Chemical Modification and Protein Stabilization

Mechanisms of chemical stabilization of proteins have been categorized under four headings:[7]

1. Crosslinking (either intra- or inter-molecular) by bifunctional reagents;
2. Strengthening of hydrophobic interactions by nonpolar reagents;

3. Introduction of new polar or charged groups, leading to additional ionic or hydrogen bonds;
4. Hydrophilization of the protein surface to reduce unfavorable surface hydrophobic contacts with water.

To this list can be added protein coupling to polysaccharides (or to polyethylene glycol) and bilayer encagement.

Crosslinking Studies

Many examples of protein stabilization by crosslinking have been reported using a wide range of 'bridging' reagents. Glutaraldehyde has been used to crosslink tetrameric lactate dehydrogenase with borohydride reduction.[24] Sixty percent of the initial activity was recovered and 82% of the product was in tetrameric form. The equilibrium transition of heat deactivation increased by 5°C to 65°C while that for guanidine deactivation increased from 0.1 M to 1.0 M.

Glyceraldehyde-3-phosphate dehydrogenase from rabbit muscle has four identical subunits and has been stabilized by crosslinking with diimidoesters (otherwise known as bis-imidates).[25] These reagents conserve the positive charge of protein amino groups. Compounds of differing length were used. Maximum stabilization occurred with dimethylpimelimidate, which has 5 methylene groups. The highest recovery of enzyme activity was 35%. A monofunctional imidoester control reagent did not stabilize the enzyme. A somewhat different crosslinking procedure also led to stabilization of this enzyme. Dicarboxylic acids of varying methylene chain lengths were activated with a (carboxyl-specific) carbodiimide. The activated diacids were then reacted with the amino groups of the dehydrogenase polypeptide.[26] Recoveries of starting activities ranged from 20-40%. Gel electrophoresis demonstrated intersubunit crosslinks. Maximum thermostabilization occurred with succinic acid: at 60°C, the native enzyme's half life was 3 minutes while the succinic acid derivative's was 20 minutes.

This two-stage carbodiimide crosslinking procedure has been used in reverse on alpha-chymotrypsin.[27] In this example, carbodiimide treatment activated carboxyl groups on the *enzyme* first and intramolecular crosslinking then resulted from addition of a homologous series of aliphatic diamines. Maximum thermostabilization occurred with the butan-1,4-diamine derivative which was 3 times more stable than the native at 50°C. Further crosslinks could be introduced by succinylation of the protein to introduce extra carboxyl groups. In this case the ethan-1,2-diamine derivative proved to be 21 times more stable than the native at 50°C.

Other reports describe crosslinking of alpha-chymotrypsin via artificially-introduced thiol groups. Free -SH groups were activated with 5,5'-dithiobis(2-nitrobenzoic acid). Dithiol compounds of varying molecular length were added to portions of activated, thiolated enzyme. Formation of disulfide-mediated molecular crosslinks proceeded spontaneously.[28] Enzyme fractions modified in this way were more resistant to guanidine hydrochloride denaturation than either native or single point-modified enzyme. Optimum stabilization occurred with pentan-1,5-dithiol. This report is noteworthy for the apparent 'self-stabilization' of the treated enzyme. When the modified, activated enzyme was presented with an equimolar mixture of dithiols, the degree of stabilization increased with time of incubation until it resembled that achieved with pentan-1,5-dithiol. This suggests a gradual exchange and optimization of the stabilizing dithiol crosslinks. The 1,5-dithiol also yielded the optimum recovery of catalytic activity (55%).

Crosslinking stabilization is not confined to intensively-studied, 'ideal' enzymes like those detailed above. Enzymes with important functions, or with diagnostic significance, have also featured in reports of crosslinking stabilization. Examples include acid phosphatase,[29] hemoglobin,[30] and penicilinase.[31] Creatine kinase and alanine aminotransferase have been stabilized by use of the homobifunctional reagent dimethylsuberimidate (refs. 32 and 33 respectively), stability being estimated by accelerated storage testing.

Yang and Olsen have described the intersubunit crosslinking of the abnormal hemoglobin New York [beta113G15)Val→Glu] by bis(3,5-dibromosalicyl) fumarate. This reagent crosslinks Lys82beta$_1$ with Lys82beta$_2$ resulting in a 13.5°C increase in the unfolding T_m value from 40.2 to 53.7°C for the oxidized methaemoglobin form. Quite similar stabilizing effects were observed for normal haemoglobin A.[30] The crosslink holds alpha-beta dimers together and favors reassociation to functional tetramers. The likely molecular interactions leading to stabilization of the crosslinked hemoglobin tetramer are discussed. These few studies indicate that crosslinking can be effective for stabilization of single chain or multisubunit proteins.

Klemes and Citri undertook crosslinking of the exopenicillinase from *B. cereus* strain 569/H with a range of compounds including cyanuric chloride, hexamethylene diisocyanate, diethyl malonimidate and glutaraldehyde. All of the resulting derivatives were more stable than the native enzyme, as measured by the remaining activity after 3 minutes exposure to 56°C at pH 7.3. Substrate-induced conformational changes were lessened by crosslinking but not eliminated, and it appears that at least some degree of conformational flexibility is required for catalysis.[31]

Dimeric beef heart creatine kinase was crosslinked with dimethylsuberimidate (DMS), an amino-specific bis-imidate with an aliphatic $(-CH_2-)_6$ bridge separating the reactive imidate functions. Despite some loss of activity, stabilization occurred and was optimal when about 50% of the available amino groups were modified. Persistence of activity at 55°C was much greater for the modified form and the T_m for activity retention increased from 49° to 54°C. The DMS-treated enzyme was also much more resistant to urea- and guanidine hydrochloride-induced activity loss.[32]

DMS has also been used on alanine aminotransferase (ALT), a clinical diagnostic marker of liver function. The first-order inactivation rate constant at 37°C of the DMS-treated beef heart enzyme fraction was 4.7 times lower than that of untreated ALT. A rigorous accelerated storage experiment indicated that the treated fraction would have a 43 fold longer shelf life at 4°C. Two shorter homologs of DMS, dimethyl adipimidate and dimethyl pimelidate, with $(-CH_2-)_4$ and $(-CH_2-)_5$ aliphatic bridges respectively, gave much lower stabilization factors when used on ALT.[33] Later investigations on pig heart ALT showed that the DMS modification also benefitted ALT's T_{50}: the value for modified ALT was 10°C higher than the native enzyme's 46°C. DMS-ALT also had a greater tolerance of some organic solvents: the modified enzyme retained greater percentages of aqueous-phase activity at all concentrations of dimethylformamide, dimethylsulfoxide and 1,4-dioxane tested. While sharp threshold inactivations occurred in all cases at >40% (v/v) concentration of each solvent, the DMS-ALT consistently expressed more activity at higher concentrations than did its unmodified counterpart. In an effort to distinguish between the effects of crosslinking and of a straightforward chemical modification of ALT lysine residues with imidoesters, pig ALT was reacted with methyl acetimidate, a monofunctional imidate incapable of forming a crosslink. This actually destabilized the enzyme with respect to heat (T_{50} 42°C), but the acetimidate-modified ALT was as tolerant of 1,4-dioxane and dimethylsulfoxide as was DMS-ALT. All three ALT forms were exposed to 3 M guanidine HCl for one hour at 25°C followed by a 60-fold dilution in assay mixture and a 30 minute resting period at room temperature. At the end of this resting period, the native, acetimidate-modified and DMS-ALT fractions recovered 50%, 64% and 90% initial activity respectively.[34] These findings indicate that the stabilizing effects of DMS on ALT are not solely due to a crosslinking effect; simple chemical modification (acetylation) of ALT lysines can benefit ALT's tolerance of solvents and denaturants. DMS-modified pig heart ALT was further stabilized following immobilization via its carboxyl groups (previously activated with 1-ethyl-3-(3-dimethylamino-propyl)carbodiimide) on adipic acid hydrazide-agarose gel particles. The immobilized-modified ALT was considerably more resistant to heat (50°C), 1,4-dioxane and urea than were its native, modified-only or immobilized-only counterparts.[35] This latter report shows how different stabilizing strategies may sometimes be combined to achieve further gains in operational stability.

Pseudomonas aeruginosa produces an elastase enzyme that is markedly different from mammalian elastases. This bacterial enzyme is a zinc-containing endoproteinase with a molecular weight of 33,000 and a pI of 5.9. It is optimally active at neutral pH and its three-dimensional structure bears a remarkably close resemblance to that of thermolysin. These two proteins each have a catalytically-essential zinc atom and an active site histidine. The elastase, however, has two disulfide bridges formed from its four cysteines, whereas thermolysin completely lacks cysteines. Elastase has a proven potential in enzymatic peptide synthesis and chemical modification was undertaken in an effort to improve its stability characteristics in the organic cosolvent systems which favor the peptide product. Reaction of elastase with dimethylsuberimidate led to a two-fold increase in catalytic stability at 70°C with greater degrees of stabilization apparent at lower temperatures. Native and modified elastase shared the same T_m value of 72°C, however, indicating that no alteration in conformational stability had taken place despite the benefits to catalytic persistence.[36] (Interestingly, conjugation of the enzyme with polyethylene glycol, or PEG, gave a similar degree of catalytic stabilization;[36] PEG modification of proteins is discussed in more detail later in this chapter.)

Arnold and colleagues have undertaken a novel form of crosslinking which combines protein engineering (chapter 8) with chemical modification. Cytochrome c contains a histidine at position 39. This lies close to position 58 across a short segment of antiparallel beta-sheet. A mutant was prepared containing histidine at position 58 and a serine-for-cysteine substitution at position 102. A H39-H58 cytochrome c mutant will be able to chelate a metal ion. (The C102S substitution is to prevent dimer formation under oxidative stress conditions.) This mutant cytochrome c was next reacted with a ruthenium bipyridyl compound, $Ru^{II}(bpy)_2$, to bring about a ruthenium-mediated crosslink between H39 and H58 (Fig. 6.1). Following purification, analysis showed that the protein preparation had been modified with ruthenium at a single site and that both of the chelating histidines were coordinated to the $Ru^{II}(bpy)_2$. T_m for the Ru^{II}-crosslinked cytochrome c increased dramatically to 72.8°C, compared with a value of 49.6° for the H39H58 mutant in the absence of Ru^{II} and 56.3° for the wild type. The unfolding Gibbs energy was almost doubled for the crosslinked mutant. Construction of a second Ru^{II}-crosslinked mutant, H4H8, spanning a much smaller loop distance, showed that

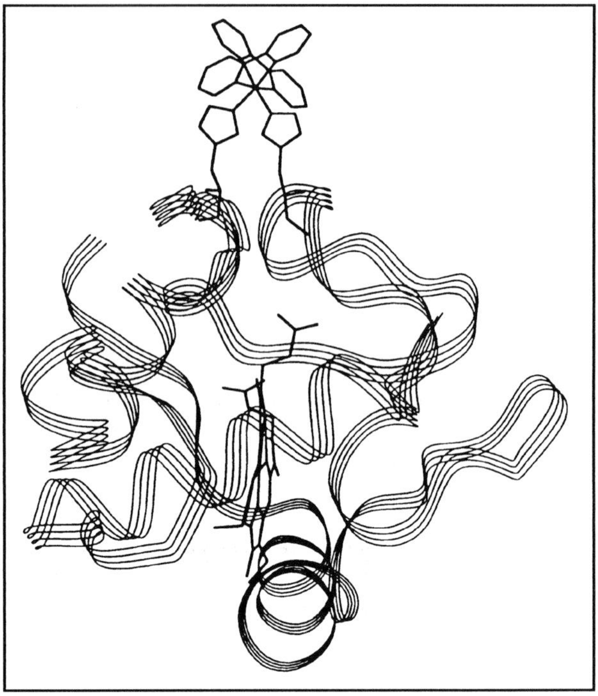

Fig. 6.1. Energy-minimized model of $Ru^{II}(bpy)_2$-$H_{39}H_{58}$ cytochrome *c*. The calculation employed Biograf, version 3.0. Coordinates for the Ru^{II} complex were taken from the structure of $Ru^{II}(bpy)_2Cl_2$. Two histidines crosslinking a beta-sheet provide an appropriate geometry for metal chelation; thus the crosslink introduces minimal changes in the backbone structure upon energy minimization. Reprinted with permission from: Muheim A, Todd RJ, Casimiro DR et al. J Am Chem Soc 1993; 115:5312. © 1993 American Chemical Society.

achievement of substantial stabilization depends on the size of the loop bridged by the crosslink, as this latter mutant was not at all stabilized.[37]

Crosslinking of Crystalline Enzymes

In an unusual approach, St Clair and Navia undertook glutaraldehyde crosslinking of thermolysin *crystals*, leading to production of crosslinked enzyme crystals (or CLECs™).[38-40] The 'art' of growing protein crystals has advanced greatly due to the requirement for high-quality protein crystals for X-ray diffraction studies of protein structures. Suitable search strategies can permit fast and efficient exploration of possible crystallization conditions, leading to reliable and reproducible crystallization protocols. Also, diffraction-quality enzyme crystals are not necessary for crosslinking-stabilization. This makes it easier to devise easy and predictable conditions for crystal growth. Indeed, microcrystals are preferable since their small size prevents undue problems of substrate or product diffusion. One can change the CLEC particle size by adjustment of the crystallization conditions.[41] Although protein crystals may break or dissolve easily, chemical crosslinking serves to stabilize both the crystal lattice and its constituent enzyme molecules.[39,41] The crystal structure is mandatory for CLEC stability: non-crystalline, crosslinked thermolysin was no more stable than the soluble enzyme. Neither was crosslinked precipitated thermolysin.[41] So why does the crystal stabilize its constituent protein molecules? It is possible that the number of protein-protein polar/electrostatic and hydrophobic interactions is greatly increased in the crystal lattice compared with a protein solution. This increased number of favorable interactions may prevent denaturation or inactivation.[41] Crosslinking, in its turn, serves to maintain the crystal lattice structure in systems other than the crystallization conditions.[41] (Note that St. Clair and Navia actually refer to crosslinked enzyme crystals as *immobilized* enzymes since the crystals are particulate and can be separated from a feedstock as simply as any conventionally-immobilized enzyme bound to a solid particle.[39] However, their use of a protein crosslinking strategy along with the absence of an exogenous solid phase has led to the description of CLECs here rather than in chapter 7.)

As its name suggests, thermolysin is a heat-stable protease with potential for peptide synthesis; it has been used to prepare the artificial sweetener aspartame (L-aspartate-L-phenylalanine methyl ester).[42] Thermolysin microcrystals were prepared by dilution in water of a special, highly-concentrated thermolysin suspension. Treatment with 12.5% glutaraldehyde effectively crosslinked the crystals which were then washed and lyophilized. The crosslinked thermolysin crystals retained virtually full activity over 5 days at 65°C and completely resisted proteolysis by added pronase. Free thermolysin completely inactivated under either of these conditions within one day. The thermolysin CLECs were also much more tolerant of pH extremes and of water-miscible organic solvents such as tetrahydrofuran.[38-40] In the type of crystal used for CLEC production, thermolysin molecules lie around a hexagonal axis of symmetry. This arrangement results in large pores throughout the crystal, aiding substrate and product diffusion. The glutaraldehyde crosslinker is thought to react with lysine residues of thermolysin to form a network of intermolecular crosslinks. However, the crosslinking is not due merely to Schiff base formation: it is irreversible even in the presence of exogenous amines.[41] Glutaraldehyde tends to polymerize when left in aqueous solution at neutral pH for an extended period. The resulting polymers possess alpha,beta-unsaturated formyl groups.[43] It seems that aqueous solutions of glutaraldehyde comprise a mixture of oligomers with differing lengths and structures; the crystal may 'choose' the most appropriate variant to form a crosslink.[41] This recalls the 'self-stabilization' noted by Torchilin et al in their dithiol crosslinking of alpha-chymotrypsin.[28] The authors argue that glutaraldehyde is the agent of choice for CLEC production: it is cheaper than many other bifunctional crosslinkers and, in certain contexts, is known to (and permitted by) regulatory authorities. This could perhaps speed up approval of products manufactured using CLEC technology.[39] However, one must optimize the crosslinking conditions to ensure CLECs of high activity and stability: typical variables include temperature, pH, reaction times and concentrations. Optimal stabilization of subtilisin Carlsberg, for instance, occurred with modification of three lysines (53% activity) while modification of more lysines resulted in further loss of activity (down to 25%).[44] (These crosslinked subtilisin crystals showed increased thermostability and decreased autolysis and remained active in dry acetonitrile.[44]) Too many crosslinks may lead to activity losses and/or to precipitation.[41]

Many other enzymes besides thermolysin have been prepared as CLECs and five CLEC enzymes (two microbial lipases, thermolysin, subtilisin and penicillin acylase) have been produced on a multi-kilogram commercial scale.[41] Already CLECs have been used for chiral compound resolution,[45] peptide synthesis[46] and other applications. The crystallization strategy has another advantage: it has meant that the CLEC presentation of *Candida rugosa* lipase (CRL-CLEC) is significantly purer than any of its commercial competitors. As a consequence of this, the CRL-CLEC is notably more stereoselective than its rival preparations.[41] Khalaf and

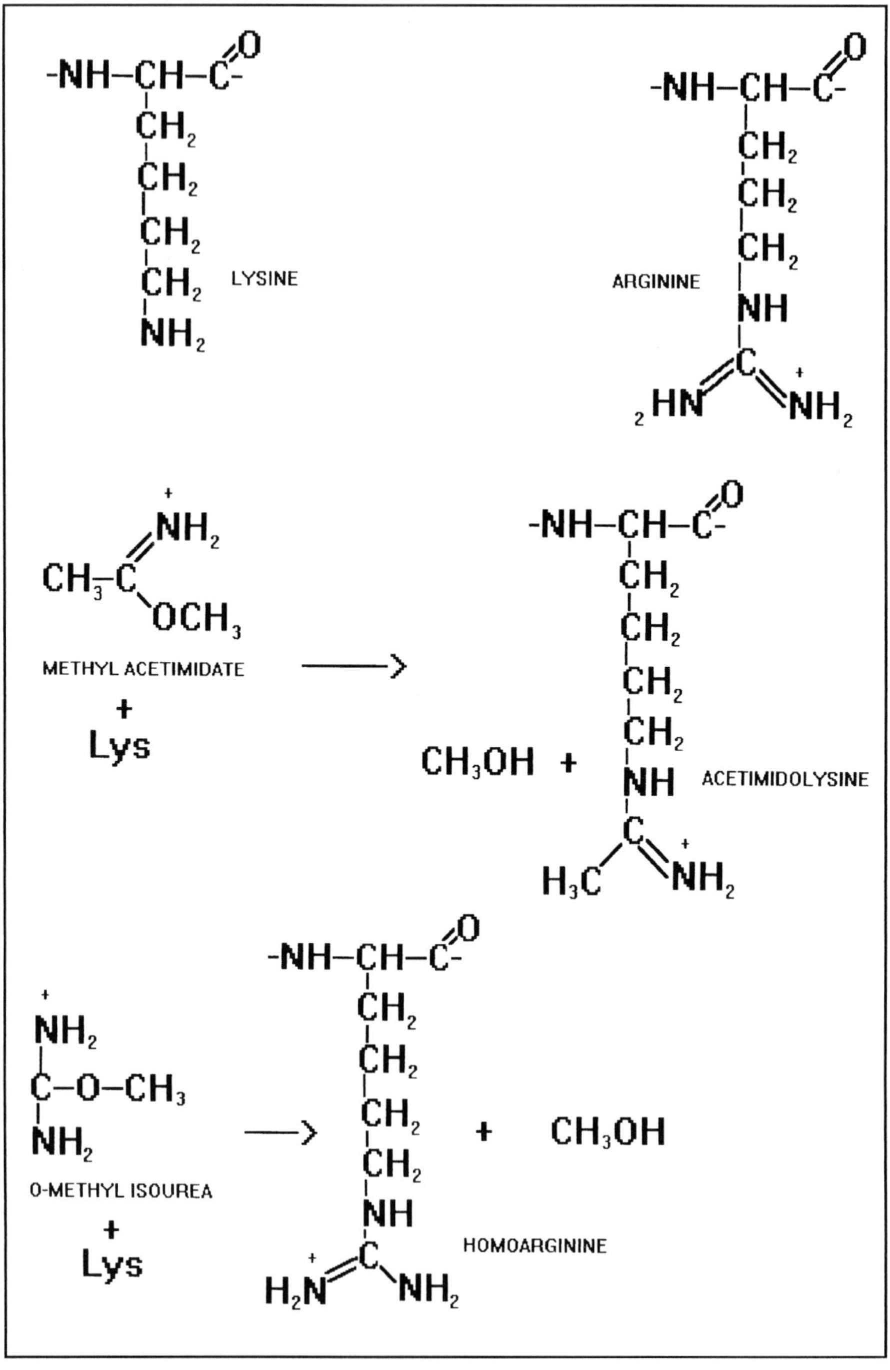

Fig. 6.2. Modification of lysine residues with methyl acetimidate and o-methyl isourea and comparison with structure of arginine.

colleagues have described a drying procedure for CLECs which yields a dry CLEC powder that is highly active in organic solvents. The drying protocol involves washing the CLECs with solutions of different surfactants in organic solvents.[47] Surfactants can increase the activity of CLECs.[41] Reference 41 provides a clear summary of the features and potential of CLECs.

New Polar or Surface Charged Groups

Crosslinking is not a prerequisite for protein stabilization by chemical modification. Dramatic increases in stability can result merely from the alteration of surface groups. Chemical modifications other than crosslinking can have dramatic stabilizing effects. Prominent among these is the introduction of new polar or charged groups onto a protein surface. The monofunctional reagent methyl acetimidate was used to alter 17 of the 24 available lysines of pig heart lactate dehydrogenase.[48] This acetamidination takes place with retention of charge but the pK values rise from 10.2 to 12.5 as the simple charged amino of the native lysine residue is replaced (Fig. 6.2). The modified enzyme was more resistant to heat, alkaline pH and even to trypsin digestion. This last finding was ascribed to the structure of the covalently-altered lysines, which is not equivalent to either lysine or arginine, the two amino acids which are cleaved by trypsin (Fig. 6.2). The increases in thermal half life ranged from 8- to 50-fold, depending on the elevated temperature used.

Charge retention is not, however, always required for stabilization by monofunctionals. Alpha-chymotrypsin has been alkylated (using acrolein followed by borohydride reduction, resulting in retention of positive charges) and acylated (with acetic or succinic anhydride, leading to the neutralization or reversal of positive charges, respectively).[49] All of these reagents react with free amino groups of the protein. Thermal stabilities increased with the degree of substitution of the available amino groups up to a maximum of 90%. Further substitution led to a dramatic drop in thermostability. The stabilizations achieved were shown to depend on the degree of modification rather than on the type of modifier used. In other words, either alkylation or acylation reactions stabilized the enzyme provided that each substituted the free amino groups on the enzyme to an equivalent degree. This suggests that the positive charges on the surface of alpha-chymotrypsin (or, at least, some of them) are not important for retention of native conformation.

Similar findings were reported for alanine aminotransferase (ALT) following reaction of its free amino groups with succinic anhydride, which reverses the positive charges of protein amino groups. This modification led to a 2-fold stabilization at 37°C (measured by comparison of first-order inactivation rate constants) and to a six-times longer shelf life at 4°C (measured by an accelerated degradation methodology). These benefits depended strongly on the use of a large excess of modifier over protein.[33]

In contrast, Cupo and colleagues observed that guanidination of chymotrypsinogen lysines by O-methyl isourea led to increased stability.[50,51] The final effect of this reaction resembles the acetamidination discussed above (ref. 48) but here a full guanidino group is added to the lysines to yield homoarginine. Net positive charge is retained. Subsequently, the remaining available lysines were neutralized by acetylation and the resulting derivative was less stable than the native form. A superguanidinated chymotrypsinogen was prepared by further guanidination of altered protein carboxyl groups and proved to be even less stable. Taken together with the results in ref. 49, these findings show just how critical the correct choice of reaction chemistries and stoichiometries can be in chemical stabilization experiments. They may also reflect the different methods used to measure stability: loss of catalytic activity[49] and hydrogen isotope in- and out-exchanges.[50,51]

Hydrophilization of Protein Surface

This has been well illustrated by the work of Mozhaev and colleagues.[52] Their approach was to 'hydrophilize' enzymes by chemical modification of non-polar surface clusters. The resulting derivative will have a greater hydrophilic surface area and be better solvated. Since its interactions with the solvent will be improved, the protein will be less likely to unfold in the face of denaturing influences.

Using alpha-chymotrypsin once again as target enzyme, dramatic stability enhancements (1,000-fold at 60°C) have been achieved for surface-hydrophilized derivatives arising from reductive alkylation of up to 10 amino groups.[52,53] Alkylation with glyoxylic acid, followed by cyanoborohydride reduction, introduced a number of $-NCH_2COO^-$ groups. These are much more hydrophilic than the naturally-occurring free $-NH_2$ groups on the protein surface and they probably lead to decreased contact between water and nonpolar clusters. Six of the fourteen lysines lie close to hydrophobic residues on the protein surface.[54] The 1,000-fold stabilization reported is the ratio of the first-order inactivation rate constants for the native and modified forms.[53] This was a remarkable result from such a simple chemical alteration with low molecular weight compounds.

Next, the enzyme surface was altered by treatment with anhydrides (or chloroanhydrides) of carboxylic acids.[52] This allows a variable number of carboxylic groups (up to 5) to be added to each lysine residue altered. Acylation of the enzyme in this way results in a more hydrophilic protein (measured by partitioning in an aqueous biphasic system) with greatly increased thermostability. None of these reagents formed a crosslink. Recoveries of activity were 30% or greater. The modified alpha-chymotrypsin was estimated to be 300 times more stable than the native by comparison of the first order inactivation constants. However, a highly unusual phenomenon was soon noticed for alpha-chymotrypsin fractions modified with pyromellitic dianhydride or glyoxylic acid, and for other hydrophilized derivatives of alpha-chymotrypsin and trypsin.[55] Arrhenius plots (chapter 1) of the rate constant of irreversible inactivation yielded a zig-zag or sawtooth pattern in place of the customary linear plot of negative slope. The thermoinactivation rate constant increased with temperature between 45-55°C, *decreased* with increasing temperature between 55-65°C and finally increased with temperature once again above 75°C (ref. 55). Neither unmodified alpha-chymotrypsin nor a derivative 'hydrophobized' by treatment with acetic aldehyde gave these anomalous Arrhenius plots. The negatively-sloping, 'normal' segments of the unusual alpha-chymotrypsin Arrhenius plots indicated apparent activation energy values of 210 $kJ.mol^{-1}$ or greater (40-50°C segment) and 105-165 $kJ.mol^{-1}$ (>70°C). A reversible conformational transition of hydrophilized alpha-chymotrypsin in the 50-60°C range was postulated to explain the anomalous Arrhenius plots. Proof for this idea emerged from fluorescence spectra monitored over the first few minutes of exposure to elevated temperatues (to distinguish reversible changes and to exclude irreversible ones). The wavelength of maximum fluorescence shifted from 332 nm to 345 nm in the 45-60°C range, indicating a conformational change and coinciding with the transition to the anomalous positive slope in the Arrhenius plot. The conformational change was shown to be reversible: the wavelength of maximum fluorescence reverted to 332 nm upon cooling to 20°C. A simple model involving 'low temperature' (E_1) and 'high temperature' (E_2) active forms of the modified alpha-chymotrypsin, with an equilibrium constant describing their transition, was developed. This led to the equation $k_{in} = k_1/(1 + K) + K.k_2/(1 + K)$ describing the zig-zag character of the anomalous Arrhenius plot. k_{in} is the experimentally-observed inactivation rate constant, k_1 and k_2 are, respectively, the irreversible inactivation rate constants of the E_1 and E_2 enzyme forms and K is the equilibrium constant for the rapid $E_1 \leftrightarrow E_2$ conformational shift. The values of k_1, k_2 and K fluctuate with temperature, leading to the zig-zag plot. Form E_1 predominates at low temperatures, where $K << 1$ and $k_{in} = k_1$. At high temperatures, E_2 accounts for all the observed activity and $k_{in} = k_2$. At temperatures corresponding to the conformational transition, K approximates to 1 and both k_1 and k_2 contribute to k_{in}.[55]

Later, Mozhaev and colleagues carried out acylation (using carboxylic acid anhydrides) and reductive alkylation (with aliphatic aldehydes) of alpha-chymotrypsin.[54] The compounds used could not form crosslinks. They obtained a wide range of 'hydrophilized' and 'hydrophobized' derivatives. The stabilized enzyme forms had identical fluorescence spectra to the native chymotrypsin, indicating that the stabilization was not due to conformational effects. Increasing stability correlated with increasing hydrophilization, whether this hydrophilization was due to the nature of the modifier itself or to the number of lysines modified by the compound in question (up to a limiting value).

Hydrophobic surface tyrosine residues of trypsin were converted to aminotyrosines in two steps. Use of tetranitromethane allowed insertion of a $-NO_2$ group into the Tyr aromatic ring and this was followed by dithionite reduction to yield aminotyrosine. The degree of stabilization increased with the number of tyrosines modified up to 4: trypsin modified to this extent was >100-fold more resistant than the native to thermoinactivation.[52] Stabilization depended exclusively on the number of aminotyrosines introduced. No conformational changes were observed and it is postulated that this alteration stabilizes the native trypsin conformation. Recovery of activity was 50% or greater.

Another modification of trypsin which has proved modestly stabilizing has been the use of acetic acid N-hydroxysucinimide ester (AANHS) to neutralize the positive charges on some of the trypsin molecule's lysine residues.[56] Approximately 8 of the 14 lysines were acetylated under mild conditions. Autolysis was much reduced, as one would expect for this lysine-modified proteinase which cleaves polypeptides specifically at Lys and Arg residues. The acetylated trypsin had a T_{50} of 51°C (versus 46°C for native trypsin) and its half life at 55°C was 8.7 minutes (4.3 minutes for native) under conditions of negligible autolysis. The modified trypsin also displayed increased organotolerance at 65°C in aqueous mixtures of acetonitrile, dimethylformamide, dimethylsulfoxide and 1,4-dioxan. (At room temperature, however, its solvent tolerance characteristics were no different from those of native.) Recovery of activity following the modification procedure was 80%, as determined by titration with the colorimetric active site reagent NPGB (*p*-nitrophenyl-*p*'guanidinobenzoate); the presence or absence of the competitive inhibitor benzamidine during the acetylation reaction had little effect on the recovery figure. The amidase activity of modified trypsin was increased by 30% over that of the native.[56] Kinetic analysis revealed that the acetylated enzyme had decreased K_m and increased k_{cat} values for both amide and ester substrates. Also, k_{cat}/K_m increased for each substrate following modification and modified trypsin showed enhanced esterase activity in hydrophilic solvents compared with native enzyme. In kinetically controlled dipeptide synthesis experiments in 95% acetonitrile, the acetylated enzyme gave a 90% yield of the dipeptide N-benzoyl-L-arginyl-L-leucinamide compared with 59% for native trypsin.[57] Thus, the acetylation of trypsin with AANHS resulted in some stability gains and also in altered kinetic properties which proved beneficial for enzymatic peptide synthesis.

Tsuji has described the reductive alkylation of trypsin resulting in enhanced stability and catalytic properties.[58] Trypsin was adsorbed onto benzamidine-Sepharose™ to protect the active site and to prevent drastic surface changes. It was then treated with n-octanal and sodium borohydride. After incubation at 40°C for one hour, activities of the modified form were significantly higher than those of the native trypsin in both emulsion and solution assay systems.

PEG and Carbohydrate Coupling

There is another form of chemical modification which can benefit stability, namely the covalent attachment of large molecular weight polyhydroxy entities such as PEG (polyethylene glycol) to proteins. Coupling of PEG to proteins has been used to alter their immunogenicity and to prolong their clearance times post-injection.[59] PEG coupling can also improve enzymes' solubilities and activities in organic solvents.[60-62] PEG has been used to dissolve enzymes in non-aqueous solvents to open up a whole new area of enzyme chemistry. It can also lead to increased stability.[5,60] The chemistry involved is simple and protocols are described in ref. 5. PEG is activated by cyanuric chloride which then reacts with free amino groups on the protein surface. Alternative chemistries are possible,[59] including 1,1'-carbonyldiimidazole (requiring long reaction times),[63,64] succinimidyl active ester, phenylchloroformate and tresyl chloride methods.[64] All of these protocols exploit the nucleophilic nature of lysine residues in proteins.[64]

Lipase from *Pseudomonas fluorescens* was coupled to PEG and the modified enzyme dissolved successfully in benzene. Storage stability in benzene was good: 40% of the day-zero ester synthesis activity remained after 140 days. The optimum temperature for ester exchange was 70°C, much higher than that for ester hydrolysis by native lipase in aqueous emulsions.[60]

Trypsin's thermostability increased on attachment of PEG and the degree of thermostabilization increased with the molecular mass of the PEG moiety. Inactivation at 60°C was not first-order, so half lives could not be estimated. Over a 30-minute experimental period at increasing incubation temperatures, $PEG_{5,000}$-coupled trypsin retained about 50% activity at 80°C while the native inactivated completely at 70°C. The $PEG_{5,000}$ derivative was also more resistant to the inactivating effects of the detergent sodium dodecyl sulphate.[65]

Yang and colleagues have described the stabilization of subtilisin Carlsberg, another proteolytic enzyme, by the attachment of $PEG_{5,000}$ (ref. 66). Conjugation with subtilisin followed activation of PEG by either cyanuric chloride or nitrophenol carbonate. Both types of PEG-modified subtilisin were soluble in benzene, toluene, chloroform and tetrachloromethane. Little activity was lost in either modification protocol: recoveries of starting activity were 88% and 93-95% for the cyanuric chloride-activated (CC-PEG) and the nitrophenol carbonate-activated PEG (NPC-PEG), respectively. Activity masurements (using the succinyl-L-alanyl-L-alanyl-L-prolyl-Lphenylalanyl-*p*-nitroanilide, or sAAPF-*p*NA, colorimetric assay) indicated that both PEG-conjugated fractions were much more tolerant of heat (range 30°-60°C) and pH (range 6.5-9.5, 30°C). At 30°C, the NPC-PEG enzyme was much more stable than its CC-PEG counterpart; both were considerably more stable than the native. The two PEG-conjugated subtilisins showed similar stabilities at higher temperatures (40°C and above), however. The half lives of modified forms ranged (according to the temperature) from 3 to 28 times greater than that of native subtilisin. Over the range of pH values, both PEG-conjugated subtilisins had much greater half lives than the native but the NPC-PEG derivative had a longer half life than the CC-PEG subtilisin at all pH values investigated. Differential scanning calorimetric measurements of the respective proteins' T_m values, however, revealed no differences between the native and PEG-modified enzymes, although the beginning of the thermal transition of both modified forms was delayed by 10°C. Inactivation kinetic experiments suggested that the improved stability characteristics arose from a lessened rate of autolysis and not from increased molecular stability. Half lives of all three subtilisins decreased as pH increased; both activity and autolysis of subtilisin increase with pH. Note that one can incorporate the PEG-modified subtilisins into polymers such as polyacrylates so as to achieve further stability gains for the PEG-modified subtilisin.[67]

Soluble carbohydrates and polysaccharides may be coupled to proteins by the cyanogen bromide method. This activates sugar hydroxyl groups for coupling to protein amino groups. This has been suggested as a general method for enzyme stabilization: carbohydrate coupling helps protect against proteolytic inactivation (probably by steric hindrance) in addition to its enhancement of protein thermostability.[68] Trypsin and alpha- and beta-amylases have been coupled to soluble dextran. All derivatives were more stable than their untreated counterparts. For example, the 60°C half life of alpha-amylase increased from 2.5 to 63 minutes following modification, a 25-fold stabilization. The modified trypsin was less prone to autolysis. An increased degree of hydration may also be an important factor in the stabilization.[69] Srivastava has reported the stabilization of a Bacillus alpha-amylase to heat, urea and proteolysis following attachment of dextran and two of its derivatives. Molecular weights of the dextran fractions used were 17,700, 40,000 and 90,000. Enzyme-carbohydrate coupling took place via two activation reactions, cyanogen bromide and sodium metaperiodate oxidation. Of the two, the cyanogen bromide method yielded the amylase with better recovery (77% versus 68%), higher degree of modification (60% versus 45%) and better overall stability characteristics. Even in the absence 6calcium and starch, the dextran-coupled amylase remained 50% active after 20 minutes at 100°C. Stability was much greater when calcium and starch were present. Native amylase was almost completely inactivated within 10 minutes at 100°C. Conjugated amylase retained 31% activity after 10 minutes in 6 M urea; native enzyme completely lost activity after 5 minutes. Next, two derivatives, acetylated (hydrophilic) and methylated (hydrophobic) dextran, were coupled to amylase in addition to the base polysaccharide. These modified dextrans also stabilized amylase but the hydrophobic methylated dextran was less effective than the hydrophilic acetylated dextran which, in turn, was slightly less stabilizing than unmodified dextran. All dextran-modified fractions were much more resistant to proteinase K digestion than native amylase. The dextran-induced stabilization, taken together with the lesser effect of the more hydrophobic methylated dextran, suggests that the gains in stability might arise from the increased hydration of the polypeptide afforded by the polysaccharide. To test this, Srivastava compared the effects of millimolar and submolar levels of $MgSO_4$ and KCl on the thermal stabilities of native and conjugated amylase at 90°C. $MgSO_4$ is a high water-binding salt which will tend to dehydrate the amylase, while KCl is a low water-binding salt which should have little effect on the protein. $MgSO_4$ adversely affected both conjugated and unconjugated amylase while 25 mM KCl had very little effect. While 0.8 M KCl was destabilizing, it was much less so than 25 mM $MgSO_4$. These findings strongly suggest that the dextran-induced stabilization arises from a hydration effect.[70]

Conditions for chemical reduction proved important in the coupling of a purified *Aspergillus niger* cellobiase (beta glucosidase) to periodate-oxidized dextran.[71] Sodium cyanoborohydride was the most effective reducing agent in terms of resistance to heat inactivation at 70°C and also in terms of recovered activity. Thermostability correlated with the number of links created between the enzyme and the soluble solid phase. Formation of the greatest number of links (15) required a reaction time of up to 24 hours. Thermostability increased with the number of lysines modified. It was thought that the increased stability arose from rigidification of the enzyme's three-dimensional structure. Beta-amylase and pullulanase have likewise been stabilized by polysaccharide attachment.[72]

Stability of a *Bacillus* sp. alkaline serine proteinase (ALP I) benefitted from conjugation with a dialdehyde derivative of clinical dextran. ALP I, a member of the subtilisin superfamily, is unstable at 50°C and is prone to autolysis. (The dialdehyde dextran had been prepared using sodium periodate oxidation to introduce aldehyde functions at the C-2 and C-4 positions of glucose residues within the dextran polysaccharide.) Following purification by column chromatography, the modified ALP I contained 75% carbohydrate by weight and 60% of its amino groups had been modified. Since ALP I contains four lysines and an N-terminus, it is likely that three amino groups were modified. Specific activity against casein decreased to 56% of native value after modification and the enzyme's pI also decreased from 8.2 to 5.0. The dextranized ALP I was 10-15°C more thermostable, depending on the temperature, at pH 10.0 over a 10 minute period, and more tolerant of anionic surfactants than was the native. Conjugated ALP I retained virtually full activity over 48 hours in 0.03% sodium dodecyl sulfonate while the native displayed very little residual activity. The dextranized enzyme was also twice as active as native ALP I in the hydrolysis of the fluorigenic peptide substrate succinyl-L-alanyl-L-alanyl-L-prolyl-L-phenylalanyl-4-methylcoumaryl-7-amide in benzene and in toluene. (However, all of the activities observed in organic solvents were less than those in aqueous buffer.)[73]

In spite of the successes outlined above, one should note that carbohydrate coupling does not always result in stability gains. Longo and colleagues did not observe any notable changes in the thermal stability characteristics of lysozyme following the attachment of cyanogen bromide-activated sucrose molecules to the polypeptide. A levansucrase enzyme from *Bacillus subtilis* was then used to add further sucrose residues to the chemically-coupled sucrose moieties on the enzyme. The failure of carbohydrate conjugation to stabilize the target protein was ascribed to the intrinsic hydrophobicity and stability of the lysozyme polypeptide.[74]

Other Strategies

A novel procedure for the stabilization of soluble enzymes has been demonstrated.[75] Termed 'encagement', it involves the formation of a crosslinked bilayer around the enzyme molecule. The enzyme's freedom of movement is greatly constrained by this net or cage and it thus becomes more resistant to denaturation. The encagement process is quite simple. One prepares a low molecular weight glutaraldehyde polymer and uses this to form a monolayer of polymeric glutaraldehyde around the enzyme. Next, one treats the enzyme with polyacrylamide derivatives bearing primary amine or acyl hydrazide groups to form crosslinks. In this way, a bilayered crosslinked cage (or net) constrains the enzyme's conformational flexibility and thus stabilizes it. One can then immobilize the bilayer-encaged enzyme within a polyacrylamide gel to achieve further stabilization. This procedure has been used for glucose oxidase, pencillinase and pig liver carboxylesterase (see ref. 75 for a detailed description).

Finally, Shami and colleagues have described an exquisitely specific form of 'chemical modification' for stabilization. This involves the binding of specific antibodies to the target protein to prevent thermal unfolding or proteolysis. Following optimization of the enzyme-antibody ratio, half lives for the catalytic activities of (polyclonal) antibody complexes of alpha-amylase, glucoamylase and subtilisin increased from minutes to hours at elevated temperatures.[76] Later studies involved the binding of monoclonal antibodies (individually and in mixtures) to L-asparaginase.[77] Other adverse influences tested included freeze-thaw cycles and lyophilization (chapter 9) on alpha-amylase, oxidation (sodium hypochlorite, NaOCl, on subtilisin) and ethanol (on glucoamylase).[77] Half lives for these three enzymes at temperatures in the range 65-70°C increased at least 60-fold. Unprotected glucoamylase had a 65°C half life of only 3 minutes, while that of antibody-bound glucoamylase was 3 hours. Quite similar results were obtained for subtilisin. Alpha-amylase, however, recorded a 240-fold increase in its 70°C half life from 4 minutes to 16 hours with antibody protection. In temperature profile tests designed to assess the temperature value leading to 50% inactivation (T_{50}) after 5 minutes, notable increases were observed. With antibody protection, the T_{50} for subtilisin shifted upwards by 6° from 63°C while alpha-amylase's increased by 20.5° from 66.5°C.[77,78] Unprotected alpha-amylase lost 65% activity following eight freeze/thaw cycles, while the antibody-bound enzyme retained virtually full activity. In lyophilization tests, unprotected alpha-amylase was completely inactivated; in the presence of antibody, 15% activity remained and this gradually increased to 25% 24 hours after rehydration. Antibody-complexed subtilisin was much more resistant to hypochlorite (0.05%) oxidation: it retained 81% activity after a 30 minute exposure, while free subtilisin's activity declined to 25%. Here, the protected enzyme suffered an initial 19% loss of activity; thereafter, its activity remained steady. In contrast, unprotected subtilisin suffered a continuous decline. Glucoamylase incubated for 25 hours in 2.5% ethanol suffers a 90% activity loss but the antibody-protected enzyme lost only 2% activity under the same conditions.[77,78] Note that significant stabilizations occurred not just at high temperatures but also under industrially-relevant conditions (oxidation resistance of subtilisin and ethanol tolerance of glucoamylase). The stabilizing effects observed may arise from more than one cause. The antibody may form non-covalent crosslinks on the enzyme surface, increasing its resistance to unfolding, or else the antibody may shield sites on the enzyme surface which are vulnerable to some inactivating process (particularly proteolysis).[77] Shami and colleagues note that typical antigen-antibody binding affinities are about 10^8 M^{-1}, suggesting a Gibbs energy reduction for the antigen (here, the enzyme to be stabilized) in the region of 10 $kcal.mol^{-1}$. This magnitude of Gibbs energy change falls within the 5-15 $kcal.mol^{-1}$ range typically reckoned as the energetic difference between folded and unfolded proteins. Thus antibody binding could well give increased thermodynamic stability to a target protein.[77]

This could be a very useful stabilizing strategy where one is lucky enough to have suitable monoclonal antibody (or antibodies). One would simply have to mix appropriate amounts of the antibody and of the protein to be stabilized. The antibody's high specificity and affinity would ensure binding to the target protein molecules. Thus, routine manipulations would be minimal, following initial protocol optimization. Hybridoma technology, meanwhile, would permit harvesting of the critical antibody on a repeat and permanent basis. (This would not be possible

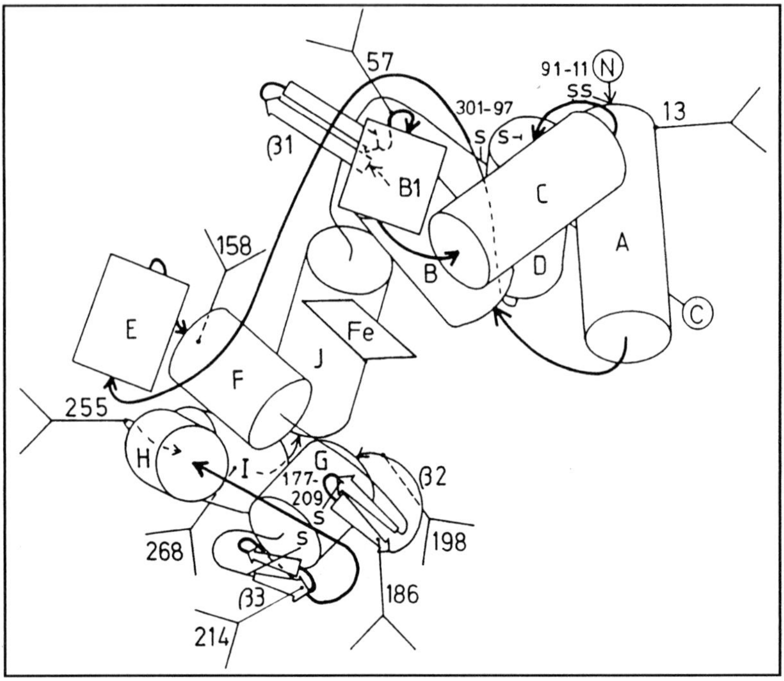

Fig. 6.3. Model structure of horseradish peroxidase, based on x-ray crystal structure of cytochrome c peroxidase. Helices A-J are drawn as cylindrical projections (diameter = 1 nm, length = number of helical residues x 0.15 nm). Extended structures, beta1, beta2 and beta3, are shown as arrows (diameter = 0.425 nm, length = number of extended residues x 0.33 nm). Reverse turns and coils are shown as connecting lines. The two domains of the peroxidase enclosing the central hemin prosthetic group (Fe in a rectangle) are pulled apart for the reason of simplicity. The amino and carboxy terminals are indicated by N and C, respectively. Three disulfide bridges of plant peroxidases are indicated by -SS- and by the residue numbers of the corresponding half-cystines of horseradish peroxidase HRP C. The eight asparagine residues of HRP C attached to carbohydrate are indicated by numbers and a symbolic carbohydrate fork. Reprinted with permission from: Welinder KG. Eur J Biochem 1985; 151:500. © 1985 European Journal of Biochemistry.

for a polyclonal antibody cocktail, no matter how stabilizing.) Raising antibodies of the required specificity would likely be costly and time-consuming, however. Quite a degree of screening would be required, since many antibodies raised against a target protein will not be stabilizing: some antibodies could block the active (or equivalent) site partially or completely, adversely affecting function, while others may bind to regions of the target protein surface with no influence on the target molecule's stability. The antibodies themselves could also have attendant stability problems.

Case Study: Horseradish Peroxidase

Horseradish peroxidase (HRP) has long been an invaluable tool in the life sciences. It functions as an indicator in oxidase-based coupled enzyme assays, in enzymeimmunoassay,[79] cytochemistry[80] and DNA probes.[81] Although it acts on a narrow range of peroxides, HRP can use a wide range of hydrogen donors including a range of chromogenic and luminescent compounds.[79,82-85] Its ability to form easily detectable compounds like these has contributed to its widespread use as a detection or probe system. It has great potential for use in biosensor configurations[86,87] and it can play a useful role in certain organic syntheses.[88-90] Its activity in organic solvents[90-93] should enhance its usefulness in these applications. Possible environmental uses include the degradation of phenols.[94]

The enzyme itself is a two-domain, hemin-containing glycoprotein weighing 44 kDa. The predominant isoenzyme C is the best-characterized form. This has a blocked N-terminus, contains four disulfide bridges and is glycosylated at up to eight asparagines. Each molecule binds two calcium ions. Its full amino acid sequence has been ascertained[95] and its carbohydrate composition has been determined.[96] It has a high turnover number and its complex reaction kinetics have been reviewed.[97] Site-directed mutagenesis is currently being used to probe its reaction mechanism[98,99] following cloning and expression of the HRP gene.[100,101] Welinder[102] has proposed a model structure for HRP based on the crystal structure of cytochrome c peroxidase (Fig. 6.3).

HRP has good stability characteristics,[97] contributing to its widespread use. This stability is likely due to its tightly bound calcium ions, extensive glycosylation at up to 8 different sites and to its 4 disulfide bridges.[95] Any worthwhile biosensor or bioreactor involving HRP, however, must be robust and stable in operation, whether in an aqueous or organic system. The enzyme must function reliably under varying conditions over an extended period. Its intrinsic stability characteristics may not be sufficient to cope with these demands. For these and related reasons, HRP has been subjected to many different types of chemical modification. These include crosslinking,[103,104] surface modification,[105,106] and attachment of PEG[107] and carbohydrate.[108] Some stability studies have been undertaken on chemically-deglycosylated forms of HRP[107,108] and on the unglycosylated recombinant enzyme expressed in *E. coli*.[109] Here, the effects of some of these procedures on HRP's stability characteristics are reviewed.

HRP Denaturation and Inactivation

Thermal inactivation of Ryan et al's commercial preparation deviated from a first-order decay at 75°C.[103] Chang et al found that HRP's thermal inactivation profile in liquid phase did not follow first-order kinetics over the range 60-94°C. They estimated the process to be 1.5 order.[110] Hendrickx et al observed a biphasic inactivation profile for solid state HRP at 140-160°C. They ascribed this to the occurrence of two fractions of HRP with independent first order kinetics.[111] Ugarova and coworkers, in contrast, reported first order thermal decay of HRP derivatives at 56°C[106] but most of their study was undertaken on the purified isoenzyme C rather than on commercial preparations. Even here, however, the native HRP log plot showed distinct curvature. This complex thermal decay suggests that procedures such as accelerated storage[112] or determination of conformational stability[113] are not appropriate for HRP since they require a two-state, first-order transition

Table 6.2. Inactivation rate constants and apparent half-lives for horseradish peroxidase and derivatives at 75°C

	Suberimidate-modified					
	Day 0			Day 4, Reduced		
	$t_{1/2}$ insp.	"k"	app.$t_{1/2}$	$t_{1/2}$ insp.	"k"	app.$t_{1/2}$
Native	3	.26 +/- .01	2.6	3	.25 +/- .04	2.8
0.1 DMS	3	.17 +/- .03	4.1	3	.22 +/- .04	3.2
0.5 "	9	.07 +/- .01	10.7	9	.07 +/- .01	9.5
1.0 "	4	.14 +/- .03	4.8	8	.07 +/- .02	9.4

	Succinimides modified		
	$t_{1/2}$ insp.	"k"	app. $t_{1/2}$
Native	5	.112 +/- .018	6
SA-NHS	43	.0174 +/- .0007	40
EG-NHS	141	.0047 +/- .0003	147

DMS = dimethyl suberimidate, concentrations in $mg.ml^{-1}$
SA-NHS = suberic acid bis(N-hydroxy succinimide ester)
EG-NHS = ethylene glycol bis(succinic acid N-hydroxy succinimide ester)
$t_{1/2}$ insp. = apparent half-lives estimated by inspection of plots of % initial activity v. time
"k" = apparent first-order rate constants determined by Enzfitter software
app. $t_{1/2}$ = apparent half lives calculated from "k" values above.
Reprinted with kind permission from: Ryan O, Smyth MR, Ó Fágáin C. Thermostabilized chemical derivatives of horseradish peroxidase. Enzyme Microb Technol 1994; 16:502. © 1994 Elsevier Science - NL, Sara Burgerhartstraat 25, 1055 KV Amsterdam, The Netherlands.

(chapter 1). Pappa and Cass used guanidine hydrochloride to unfold both apo- and holo-HRP and noted changes in the enzymes' fluorescence and CD characteristics.[114] Fluorescence increased at denaturant concentrations greater than 1M and the denaturation of holo-HRP showed two distinct steps. These were attributed to unfolding of the polypeptide and to loss of the heme prosthetic group. Pappa and Cass also achieved some recovery of activity on renaturation, i.e. the unfolding was (to some extent) reversible upon removal of the denaturant. This is of some interest, as recombinant HRP expressed in *E. coli* must be renatured from inclusion bodies.[101] Inclusion of EDTA in the guanidine hydrochloride solutions permits removal of calcium ions which help maintain HRP's conformation.[115] Ryu and Dordick used fluorescence to follow inactivation of HRP by water-miscible organic solvents (acetonitrile, dioxane and methanol)[116] while Gorman and Dordick have described the desorption of HRP-bound tritiated water (T_2O) in dimethylformamide (DMF), methanol and tetrahydrofuran (THF).[117] T_2O desorption was very rapid, with most of the desorbable T_2O released in the first five minutes. HRP retained less T_2O in moderate-polarity solvents (such as DMF and THF) than did chymotrypsin.[117]

Egorov et al[109] found Chiswell and Ortlepp's[100] *E. coli*-expressed, unglycosylated, recombinant HRP to be 3-fold less stable at 56°C than commercial plant (glycosylated) enzyme. Bonnaffe et al[108] found that periodate-deglycosylated plant HRP was 2.8 times less stable at 50°C than untreated enzyme while Hiner et al[118] noted that carbohydrate-free recombinant HRP C* was twice as sensitive as plant HRP to H_2O_2 inactivation.

Chemical Modification of HRP

Ryan et al used bis-imidates in an attempt to form stabilizing intramolecular crosslinks in HRP. Imidoesters react with protein amino groups, for which they are quite specific.[103] The bifunctional imidoesters used were dimethyl adipimidate (DMA), dimethyl pimelimidate (DMP) and dimethyl suberimidate (DMS). These have respectively 4, 5 and 6 methylene groups between the reactive end groups. Modification with the monofunctional methyl acetimidate was also used, so as to distinguish between the effects of crosslinking and a straightforward chemical modification.

Treatment with imidoesters led to noticeable increases in HRP's thermal stability at 75°C. Table 6.2 shows apparent half lives on the day of modification for HRP treated with various concentrations of DMS. Higher concentrations of DMS, 0.5 and 1.0 $mg.ml^{-1}$, led to increases in the 75°C half life, that is, to a more stable HRP. The 0.5 $mg.ml^{-1}$ concentration seemed to be optimal, as concentrations above this did not lead to any further stabilization. When these modified HRPs were stored for a few days at 4°C and then tested again for thermal stability at 75°C, no enhanced stability was seen. When DMS-treated HRP fractions were reduced with borohydride, however, and compared with unreduced fractions, the initially observed increases in thermal stability were maintained on storage at 4°C, at least up to 4 days. Table 6.2 includes apparent half lives for modified and reduced HRP fractions after 4 days of storage and shows that the initial stabilizations following DMS treatment were maintained. The near-doubling of the 1.0 $mg.ml^{-1}$ DMS fraction's apparent half life between 0 and 4 days is probably not significant when the errors in the "k" values are taken into account. When DMA and DMP were used in place of DMS, transient stabilization of HRP was observed at certain concentrations. These did not persist on storage, however, and borohydride reduction was not successful in 'locking' the initially stabilized forms, in contrast to its beneficial effect in the case of DMS-treated HRP. The monofunctional imidoester methyl acetimidate was also used to modify HRP under conditions similar to those used for the bis-imidoesters. Transient thermal stabilization was noted at certain concentrations but did not persist on storage. Borohydride reduction did not maintain the initial stabilization. In these respects methyl acetimidate resembles the DMA and DMP bis-imidoesters. No losses of initial HRP activity were observed following imidoester treatment.

Next, a different class of amino-specific reagent, the N-hydroxy succinimide (NHS) esters, was used to see if extra stabilization of HRP could be achieved. Two such homobifunctional compounds, suberic acid bis(N-hydroxysuccinimide ester) (or SANHS for short) and ethylene glycol bis(succinic acid N-hydroxysucinimide ester) (abbreviated as EGNHS) were reacted with HRP. Both bis-succinimides yielded greatly stabilized derivatives without any loss of HRP activity. Table 6.2 shows the apparent half lives of native and NHS-modified HRPs. Stabilizations of 6- to 23 fold were observed, depending on the NHS ester used. These stabilizations persisted on storage, unlike those induced by the imidoesters.

A partial characterization of the NHS derivatives indicated no molecular weight differences between modified and unmodified HRPs. Thus it appears that the NHS

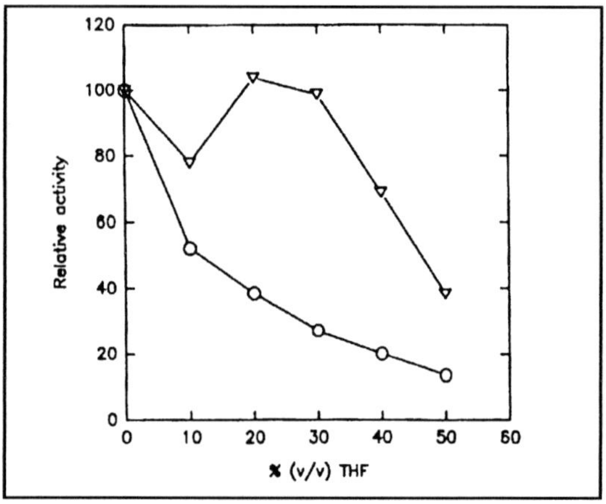

Fig. 6.4. Exposure of horseradish peroxidase forms to tetrahydrofuran/buffer mixtures for 60 minutes at 25°C. Native HRP, ○; Acetylated HRP, ●. Buffer was 0.01M phosphate, pH7.0. Reprinted with kind permission from: Miland E, Smyth MR, Ó Fágáin C. Enzyme Microb Technol 1996; 19:65. Increased thermal and solvent tolerance of acetylated horseradish peroxidase. © 1996 Elsevier Science - NL, Sara Burgerhartstraat 25, 1055 KV Amsterdam, The Netherlands.

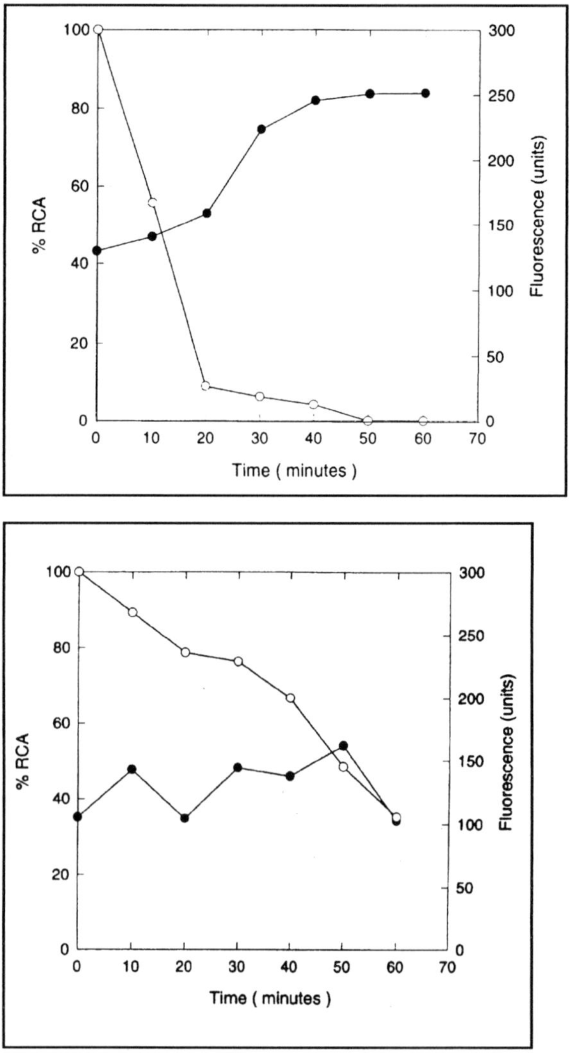

Fig. 6.5. Remaining catalytic activity (rca)/fluorescence profile of native (a, right) and egnhs-modified horseradish peroxidase (b, below) at 65°C. HRP activity, ○; fluorimetric emission, ≅. Reprinted with kind permission from: Miland E, Smyth MR & Ó Fágáin C. Modification of horseradish peroxidase with bifunctional N-hydroxysuccinimide esters: effects on molecular stability. Enzyme Microb Technol 1996; 19:247. © 1996 Elsevier Science - NL, Sara Burgerhartstraat 25, 1055 KV Amster dam, The Netherlands.

ester stabilizations took place without intermolecular crosslinking or oligomerization of the HRP. The number of free amino groups on native and modified HRP was ascertained; only one TNBS-reactive lysine epsilon-amino group remained after reaction with SANHS or EGNHS. Modification of all six HRP lysines seems to be difficult. Ugarova et al did not achieve modification of all six lysine amino groups with the same modifier when using a variety of anhydride compounds. They noted that TNBS reacted with all six aminos only at 40°C; the same reaction at 4°C modified only three of them. Reaction of HRP with anhydrides at 0°C modified between three and four amino groups.[106] Lys174 interacts with the prosthetic heme group[119] and so is less likely to react with chemical modifiers. Experiments aimed at optimizing the protocol for modification with EGNHS alone revealed that the degree of stabilization increased with modifier concentration up to 4-5 $mg.ml^{-1}$ and that the optimal reaction time was 20 minutes.[104] There was no significant difference between the pH-activity profiles of native and NHS-treated HRPs in the range 5.0-8.5.[103]

Miland et al explored this modified form's resistance to adverse influences other than heat, notably water-miscible organic solvents and guanidine hydrochloride. The solvents used were methanol, dimethylformamide (DMF) and tetrahydrofuran (THF): these have respectively low, medium and high denaturation capacities (chapter 3 and ref. 120). Native and EGNHS HRPs were exposed to increasing 10% volume increments of the three solvents in pH 7 buffer, either for 60 minutes at room temperature or for 10 minutes at 60°C. There was little difference between the native and modified HRPs' resistance to methanol at room temperature. Unlike native HRP, however, both modified forms tolerated up to 20% methanol at 60°C; indeed, they showed a 2-fold activation. (Vazquez-Duhalt and colleagues noted a decreased apparent K_m for polyethylene glycol-modified HRP in methanol.[107]) DMF had quite severe effects: none of the HRP derivatives had significant activity at 50% DMF, irrespective of the temperature. The modified HRPs resisted DMF to a much greater degree at both temperatures, however. The modified HRPs also resisted THF to a greater degree, at either temperature, than did the native enzyme. Curiously, however, native HRP was much more tolerant of THF than of DMF at room temperature (Fig. 6.4), despite THF's higher denaturation capacity.[120] No significant activation effects were observed with either of these solvents. Miland and colleagues also correlated activity loss at 65°C with fluorescence changes in the native and modified enzymes.[104] About 90% of native HRP activity was lost within 20 minutes, *before* a near-doubling of fluorescence intensity occurred between 20-30 minutes. Ryu and Dordick followed changes in HRP's 403 nm Soret band absorbance peak (i.e. in its heme environment) at 65°C, noting little change up to 20 minutes and a decrease in heme absorbance after this time. Possibly, quite a small conformational change leads to loss of the heme, giving an inactive but still-folded apoenzyme. The increased fluorescence after 20 minutes suggests that the HRP polypeptide undergoes thermal denaturation. The EGNHS-stabilized enzyme showed little net change in fluorescence intensity over the period of experiment, although there are variations between individual time points (Fig. 6.5). EGNHS-HRP was also more resistant to denaturation by 6 M ganidine hydrochloride-10 mM EDTA, suggesting the existence of a crosslink. (A modified but 'uncrosslinked' polypeptide would be unlikely to resist unfolding by guanidine hydrochloride.) Mercaptoethanol addition to guanidine hydrochloride-treated samples resulted in complete inactivation of HRP, irrespective of modification, but mercaptoethanol had no effect in the absence of guanidine hydrochloride. Thus, the four disulfide bridges are critical for HRP's molecular stability and EGNHS modification does not compensate for the loss of the -S-S- links.

Both sets of compounds used, bis-imidates and bis-succinimides, are amino group-specific and may be used under mild conditions of pH and temperature. The imidates are water-soluble and retain the positive charge of the original amino group.[18] (DMS in particular has been used to stabilize a variety of enzymes.[32-34]) Succinimides in contrast must be dissolved initially in an organic solvent. They neutralize the amino group's positive charge in the course of reaction.[18] It is interesting that such a difference should arise in the HRP-stabilizing abilities of two types of amino-specific bifunctionals. Charge neutralization may account for the greater stabilization of HRP following succinimides treatment (see next paragraph). It is highly probable that both sets of compounds have reacted with HRP's lysines. HRP's amino terminus is blocked[95] and, while succinimides can react with the imidazole ring of histidine residues,[18] the complete recovery of HRP activity after reaction shows that its essential histidines (located near the hemin prosthetic group)[95] have not been altered. Also, the glucosamine residues of the HRP carbohydrate side chains are acetylated[96] and thus unreactive. Reaction with succinimides is limited, therefore, to lysine residues of the polypeptide backbone. The carbohydrate portion (18% of HRP)[95] remains intact and can be used as a target for immobilization procedures and in biosensor construction. Of the imidoesters, only dimethyl suberimidate yielded a stabilized HRP. This crosslinker spans a distance of 11 Å. Regarding the bis-succinimides, the suberic acid derivative (SANHS) spans 11 Å while the ethylene glycol compound (EGNHS) compound is 14 Å long. The longer EGNHS stabilizes HRP to a greater degree than does SANHS. Imidoesters shorter than 11 Å failed to stabilize HRP. These results suggest that 11 Å may be the minimum distance to obtain a stabilizing crosslink. The existence of an intramolecular crosslink, however, awaits explicit demonstration.

Surface Group Modification

Treatment of HRP with a monofunctional succinimide can determine whether any stability changes might arise simply from charge neutralization oflysine residues *without* crosslinking. Miland et al showed that HRP stability is enhanced by acetylation with acetic acid N-hydroxysuccinimide ester (AANHS).[105] Half-life at 65°C increased fivefold despite this reagent's inability to form crosslinks. The apparent half lives of native and AANHS forms were 7 minutes (k = 0.098+/-0.012 min^{-1}) and 39 minutes (k = 0.018+/-0.002 min^{-1}) respectively. No loss of activity occurred during modification. AANHS treatment also resulted in greater tolerance of DMF and THF at 25°C and 60°C and of methanol at 60°C. The solvent tolerance profiles were similar to those for SANHS- and EGNHS-modified HRP. Analysis of free amino groups indicated that AANHS alters 3 HRP lysines. (Native HRP's amino group content was 0.9mM; the AANHS form's was 0.41 mM.) Thus it has proven possible to achieve a notable stabilization by means of a simple chemical modification involving neither immobilization nor the use of crosslinking agents. Why is this so?

The acetyl group introduced by the AANHS reaction is small and electrically neutral. It cannot bring about any charge reversal and is too small to provide significant shielding of close-lying groups. Neutralization of like charges' repulsion may be responsible for the stabilization observed: a decreased number of like surface charges will lessen the protein's tendency to unfold at high temperatures. There is an alternative explanation. HRP's 6 lysine residues occur at positions 65, 84, 149,

174, 232 and 241.[95,102] Lys 174 is thought to interact with the heme group;[119] it is unlikely to react with AANHS, and any such reaction would likely inactivate the enzyme. Of the 5 'available' lysines, at least three lie in regions of positive hydropathic character,[102] thereby indicating hydrophobicity. (Recall that TNBS experiments indicated modification of 3 lysines.) Neutralization of positive charges within such hydrophobic sequences may well have a thermostabilizing effect and may be relevant also to AANHS-HRP's increased 'organotolerance'. One would expect that neutralization of surface charges would improve a protein's ability to tolerate organic solvents. This seems to be the case for acetylated HRP at low concentrations of the solvents tested. The surface changes are insufficient to prevent unfolding at higher solvent concentrations, however, where the water content of the system is significantly decreased and the enzyme likely loses its bound water shell. In general, one would expect that the presence of charged residues within a hydrophobic region of a protein would be destabilizing, particularly in an organic or mixed system. Neutralization of such charges could have a stabilizing effect in a cosolvent system. Martinez and Arnold noted (in a protein engineering study) that substitution of two particular charged surface residues (Arg45 and Arg78) increased alpha-lytic protease's stability in 84% (v/v) dimethylformamide at 30°C. A total of seven substitutions were stabilizing; six of these were large, hydrophobic amino acids (see chapter 8 for further details). In contrast to the situation described here for HRP, however, their substitutions in alpha-lytic protease were not accompanied by any significant increases in aqueous thermal stability.[121] AANHS confers increased tolerance of water-*miscible* organic solvents, exactly the solvents most likely to strip away the essential water shell surrounding the enzyme.[117,122] Acetylated HRP's increased 'organotolerance' takes on added significance in the light of Batra and Gupta's report of enhanced HRP catalytic activity in DMF, THF and other solvents.[123]

Other Modifications

Other workers have described various chemical modifications of soluble (not immobilized) HRP. Ugarova et al have described the use of mono- and dicarboxylic acid anhydrides and of TNBS: these react with lysine amino groups. Some of these compounds, notably succinic anhydride, reversed the lysine's positive charge. They observed substantial stabilization of HRP catalytic activity at 56°C, accompanied by spectral changes, indicating restricted conformational mobility relative to the native. The reduced flexibility of 4 of the HRP lysines correlated with enhanced stability. The observed effects depended more on the extent of modification than on the nature of the modifier. Modification of all 6 lysines, however, actually reduced thermostability.[106]

Heterobifunctional reagents containing an amino-reactive succinimidyl group have been used to modify HRP for the preparation of immunoassay conjugates.[124,125] Many conjugations have involved reaction of the enzyme's carbohydrate side chains, such as the widely used periodate oxidation procedure for the preparation of immunoassay conjugates.[79] Arseguel and colleagues have modified the carbohydrate portions of HRP with C_8 and C_{16} aliphatic chains and have described the resulting derivatives. Some of these showed improved activity in organic solvents.[126] Urrutigoity and Souppe have described the attachment of polyethylene glycol (PEG) to periodate-oxidized carbohydrate chains of HRP. The PEG-conjugated enzyme was both soluble and active in chloroform and toluene.[127] Activity of amino-derivatized HRP in water-immiscible organic solvents has been described.

Table 6.3. Thermoinactivation of horseradish peroxidase forms at 65°C

HRP Type	k, min^{-1}	$t_{1/2}$, min	Stabilization factor
HRP C*	0.033 +/- 0.007	21	1
HRP C*, modif.	0.0064 +/- 0.0003	108	5

HRP Types: *HRP C**, recombinant horseradish peroxidase (ref. 100); *HRP C* modif.*, HRP C*, ethylene glycol bis-succinimidyl succinate-modified. Inactivation rates (k-values) are from data fit to first-order exponential decay. Half-lives ($t_{1/2}$) were calculated from $0.693/k$. Stabilization factor is the ratio of k-values.

In contrast, Wirth et al attached oxidized PEG of differing molecular weights to free amino groups of HRP rather than to its carbohydrate chains. One derivative, prepared with PEG 5000, was soluble and active in various organic solvents and was more stable in toluene than in buffer at room temperature.[128] This echoed earlier work where PEG, activated by a triazine, was coupled to HRP amino groups. The PEG-HRP conjugates were soluble and active in benzene.[93] HRP's lysine residues have also been reacted with dithioesters and the modified enzyme was soluble and active in diethyl ether.[129]

Bonnaffe and colleagues have reported stabilization of horseradish peroxidase following covalent attachment of additional carbohydrate. The extensive natural glycosylation of this important plant enzyme makes a small contribution to the observed stability (chapter 2). Bonnaffe et al attached extra carbohydrate to HRP ('surglycosylation') by the use of a substituted (succinic anhydride) derivative of glucose. Glucose as such is not a constitutent of HRP's carbohydrate moieties but attachment of 3-4 activated glucose residues to the native enzyme (presumably to lysine residues) led to a 5- to 6-fold gain in half life at 60°C without any loss of activity. The temperature of half-inactivation also increased from 60°C to 67°C following glucose coupling. Given that the glucose attachment was mediated by a charge-reversing succinic anhydride derivative similar to that used by Ugarova et al,[106] Bonnaffe and colleagues prepared a succinylated HRP to ascertain whether the observed stabilization was due to the additional carbohydrate or merely to charge reversal. The succinylated HRP's thermostability was intermediate between that of native and surglycosylated HRP; the latter was 2-fold more stable than the succinylated form, indicating that a definite stabilization had resulted from the extra carbohydrate. As few as 3 to 4 glucose residues were required to stabilize HRP, which is interesting in view of the presence in nature of >100 sugar residues[108] on the HRP polypeptide at its 8 glycosylation sites.[95,102] Bonnaffe et al suggest that the number of attached carbohydrate chains, and the presence of glucose specifically, may be determining factors in the stabilization achieved.[108]

Chemical Stabilization of Recombinant HRP

Recombinant HRP (HRP C*; ref. 101) was treated with EGNHS according to Ryan et al.[103] EGNHS-modified HRP C* lost no activity on modification and showed enhanced stability at elevated temperatures. Thermoinactivations of the various HRPs at 65°C fitted well to a first-order exponential decay. Chemical modification increased the 65°C half-life of HRP C* 5-fold (Table 6.3). Modified HRP C* tolerated 1,4-dioxan up to 60% (v/v) but not dimethylformamide (DMF). The full recovery of initial activity shows that the 'naked' HRP C* polypeptide tolerates chemical modification of its lysines with EGNHS, indicating that glycosylation does not influence HRP's ability to withstand the modification reaction without inactivation. Modification with EGNHS stabilizes HRP C*, just as it does the plant enzyme. The stabilization factor of 5-fold at 65°C is, however, lower than the 23-fold stabilization reported by Ryan et al[103] for plant HRP at 72.5°C. This may be due to the different thermoinactivation temperatures. While EGNHS-treated HRP C* showed improved tolerance of 1,4-dioxan, it was no more tolerant of DMF than was untreated HRP C*. This contrasts with Miland et al's report[105] of improved DMF tolerance following acetylation of plant HRP. Glycosylation may influence HRP's tolerance of certain solvents while having less effect on its thermal stability. It was not possible to quantitate the extent of lysine modification with trinitrobenzenesulphonic acid, due to the limiting quantities of HRP C* available. Possibly, the modifications of HRP C* were not as extensive as those of Ryan et al (ref. 103). Nevertheless, it is clear that chemical modification of HRP C* lysines significantly increases the recombinant enzyme's stability, either by preventing unfolding (of N to U; see chapter 1) or by facilitating refolding (of U to N). This may be due to a crosslinking effect (as noted above) since EGNHS is a bifunctional reagent but stabilization may also arise from succinimide-induced neutralization of lysine positive charges.[105] Nevertheless, it is clear that carbohydrate-free, recombinant HRP C* can withstand EGNHS modification of its lysines with excellent recovery of activity, and that the resulting EGNHS derivative is stabilized against heat and against 1,4-dioxane. This underlines yet again the importance of lysine residues to HRP's stability.

Conclusion

A wide range of protein and enzyme properties have successfully been altered by chemical modification procedures. These include solubility, catalytic activity, substrate selectivity and stability, the main focus of this chapter.[130] Chemical modification can dramatically increase protein stability, as shown by the numerous examples in this chapter, with good recoveries of activity. Active sites may be protected by specific ligands, if required, but some degree of inactivation is always a possibility. Partial loss of biological activity may be due either to decreased activity of all the enzyme/protein molecules in the system or to total inactivation of some of the molecules present. It can be difficult to distinguish between these alternatives. Lessened activity among the entire population of molecules often leads to altered kinetic parameters or pH profiles, while a decreased turnover number (i.e. a lesser number of active sites) accompanied by an unchanged K_m suggests the presence of a molecular fraction with unaltered activities.[130] Complete loss of biological activity following a given modification reaction is usually interpreted in terms of the targetted residue(s) being essential for activity. This is not always the case, however, even where inclusion of a substrate, coenzyme or competitive inhibitor has a protecting effect. Conformational changes within the target protein's active site (or elsewhere on the molecule) may underlie these sorts of experimental results, as pointed out in ref. 130.

Despite these reservations, chemical modifications may be carried out and characterized quite quickly. The information gained is useful for further rounds of chemical modification or for suggesting target sites for site-directed mutation of the protein of interest. The ability to create non-protein amino acid derivatives by chemical modification usefully complements the scope of mutational/protein engineering strategies for the study or manipulation of proteins.[130] Meanwhile, mutagenesis and expression techniques continue to improve while chemical techniques are likewise becoming more sophisticated. Very many reagents satisfy the main criteria of a useful protein modifier, namely a high specificity for one type of amino acid residue within a protein molecule and an ability to react with that residue under mild conditions of pH and solvent composition.[130] It is likely that these powerful and complementary strategies will be increasingly combined in the future. Indications of this have already emerged in work such as the combined mutational/chemical stabilization of cytochrome c.[37]

References

1. Lundblad RL, Noyes CM. Chemical Reagents for Protein Modification, 2 vols, Boca Raton: CRC Press Inc, 1984.
2. Means GE, Feeney RE. Chemical Modification of Proteins. San Francisco: Holden-Day, 1971.
3. Means GE, Feeney RE. Chemical modification of proteins: history and applications. Bioconjugate Chem 1990; 1:2-12.
4. Hirs CHW, Timasheff SN, eds. Enzyme Structure Part B. Methods in Enzymol 25. NY: Academic Press, 1972:387-644.
5. Imoto T, Yamada H. Chemical modification. In: Creighton TE, ed. Protein Function: a Practical Approach. Oxford:IRL Press, 1989:247-277.
6. Profy AT, Schimmel P. Complementary use of chemical modification and site-directed mutagenesis to probe structure-activity relationships in enzymes. Prog Nucl Acids Res, Mol Biol 1988; 35:1-26.
7. Mozhaev VV, Melik-Nubarov NS, Siksnis V, et al. Strategy for stabilizing enzymes. Part two: increasing enzyme stability by selective chemical modification. Biocatalysis 1990; 3:189-196.
8. Torchilin VP, Martinek K. Enzyme stabilization without carriers. Enzyme Microb Technol 1979; 1:74-82.
9. Mozhaev VV, Martinek K, Berezin IV. Structure-stability relationship in proteins: fundamental tasks and strategy for the development of stabilized enzyme catalysts for biotechnology. CRC Crit Rev Biochem Mol Biol 1988; 23:235-281.
10. Mozhaev VV, Martinek K. Structure-stability relationships in proteins: new approaches to stabilizing enzymes. Enzyme Microb Technol 1984; 6:50-59.
11. Fields R. The measurement of amino groups in proteins and peptides. Biochem J 1971; 124:581-590.
12. Fields R. The rapid determination of amino groups with TNBS. In: Hirs CHW, Timasheff SN, eds. Enzyme Structure Part B. Methods in Enzymol vol 25. NY: Academic Press, 1972:464-468.
13. Creighton T.E. Disulfide bonds between cysteine residues. In: Creighton TE, ed. Protein Structure: A Practical Approach. Oxford: IRL Press, 1989:155-167.
14. Cardemil, E. Kinetics of the chemical modification of enzymes. In: Eyzaguirre J, ed. Chemical Modification of Enzymes: Active Site Studies. Chichester, UK: Ellis-Horwood, 1987:23-34.
15. Rakitzis ET. Kinetics of protein modification reactions. Biochem J 1984; 223:259-262.
16. Eyzaguirre J, ed. Chemical Modification of Enzymes: Active Site Studies. Chichester, UK: Ellis-Horwood, 1987.
17. Shaw E. Chemical modification by active-site-directed reagents. In: Boyer PD, ed. The Enzymes, 3rd edn, vol 1. NY: Academic Press, 1970: 91-146.
18. Ji TH. Bifunctional reagents. In: Hirs CHW, Timasheff SN, eds. Methods in Enzymology 91. NY:Academic Press, 1983:580-609.

19. Kennedy JH, Kricka LJ, Wilding P. Protein-protein coupling reactions and the applications of protein conjugates. Clin Chim Acta 1976; 70:1-31.
20. Srinivasachar K, Neville DM. New protein cross-linking reagents that are cleaved by mild acid. Biochemistry 1989; 28:2501-2509.
21. Aldwin A, Nitecki D. A water-soluble, monitorable peptide and protein cross-linking agent. Anal Biochem 1987; 164: 494-501.
22. Duncan RJS, Weston PD, Wrigglesworth R. A new reagent which may be used to introduce sulfhydryl groups in to proteins, and its use in the preparation of conjugates for immunoassay. Anal Biochem 1983; 132:68-73.
23. Wong SS, Wong L-JC. Chemical crosslinking and the stabilization of proteins and enzymes. Enzyme Microb Technol 1992; 14:866-874.
24. Gottschalk N, Jaenicke R. Chemically crosslinked lactate dehydrogenase: stability and reconstitution after glutaraldehyde fixation. Biotechnol Appl Biochem 1987; 9:389-400.
25. Trubetskoy VS, Torchilin VP. Artificial and natural thermostabilization of subunit enzymes. Do they have similar mechanism? Int J Biochem 1985; 17: 661-663.
26. Torchilin VP, Trubetskoy OS, Omel'Yanenko VG et al. stabilization of subunit enzyme by intersubunit crosslinking with bifunctional reagents: studies with glyceraldehyde-3-phosphate dehydrogenase. J Molec Catal 1983; 19:291-301.
27. Torchilin VP, Maksimenko AV, Smirnov VN et al. The principles of enzyme stabilization III. The effect of the length of intra-molecular cross-linkages on thermostability of enzymes. Biochim Biophys Acta 1978; 522:277-283.
28. Torchilin VP, Maksimenko AV, Smirnov VN et al. Principles of enzyme stabilization V. The possibility of enzyme selfstabilization under the action of potentially reversible intramolecular cross-linkages of different length. Biochim Biophys Acta 1979; 568:1-10.
29. Dziembor-Gryszkiewicz E, Maksimenko AV, Torchilin VP et al. Stabilization of human prostatic acid phosphatase by cross-linking with diamines. Biochem Internat 1983; 6:627-633.
30. Yang T, Olsen KW. Thermal stability and crosslinking of Hb New York [beta113(G15)Val→Glu]. Hemoglobin 1989; 13:147-156.
31. Klemes Y, Citri N. Catalytic and conformational properties of crosslinked derivatives of penicillinase. Biochim Biophys Acta 1979; 567:401-409.
32. Sheehan H, O'Kennedy R, Kilty C. Investigation of the properties of bovine heart creatine kinase cross-linked with dimethyl suberimidate. Biochim Biophys Acta 1990; 1041:141-145.
33. Ó Fágáin C, O'Kennedy R, Kilty C. Stability of alanine aminotransferase is enhanced by chemical modification. Enzyme Microb Technol 1991; 13:234-239.
34. Moreno JM, Ó Fágáin C. Activity and stability of native and modified alanine aminotransferase in cosolvent systems and denaturants. J Molec Catal B Enzym 1997; in press.
35. Moreno JM, Ó Fágáin C. Stabilization of alanine aminotransferase by consecutive modification and immobilization. Biotechnol Lett 1996; 18:51-56.
36. Besson C, Favre-Bonvin G, Ó Fágáin C et al. Chemical derivatives of *Pseudomonas aeruginosa* elastase showing increased stability. Enzyme Microb Technol 1995; 17:877-881.
37. Muheim A, Todd RJ, Casimiro DR et al. Ruthenium-mediated cross-linking and stabilization. J Amer Chem Soc 1993; 115:5312-5313.
38. St Clair NL, Navia MA. Crosslinked enzyme crystals as robust biocatalysts. J Amer Chem Soc 1992; 114:7314-7316.
39. Navia MA, St Clair NL, Griffith JP. Crosslinked enzyme crystals (CLECs™) as immobilized enzyme particles. In: van den Tweel WJJ, Harder A, Buitelaar RM, eds. Stability and Stabilization of Enzymes. Amsterdam: Elsevier, 1993:63-73.

40. Navia MA, St Clair NL. Use of crosslinked crystals as a novel form of enzyme immobilization. Patent Cooperation Treaty Application WO 92/02617: 1992.
41. Margolin AL. Novel crystalline catalysts. Trends Biotechnol 1996; 14:223-230.
42. Gill I, López-Fandiño R, Jorba X et al. Biologically active peptides and enzymatic approaches to their production. Enzyme Microb Technol 1996; 18: 162-183.
43. Tashima T, Imai M, Kuroda Y et al. Structure of a new oligomer of glutaraldehyde produced by aldol condensation reaction. J Org Chem 1991; 56:694-697.
44. Tüchsen E, Ottesen M. Kinetic properties of subtilisin type Carlsberg in the crystalline state. Carlsberg Res Commun 1977; 42:407-420.
45. Lalonde JJ, Govardhan CP, Khalaf NK et al. Cross-linked crystals of *Candida rugosa* lipase: highly efficient catalysts for the resolution of chiral esters. J Am Chem Soc 1995; 117:6845-6852.
46. Persichetti RA, St Clair NL, Griffith JP et al. Cross-linked enzyme crystals (CLECs) of thermolysin in the synthesis of peptides. J Am Chem Soc 1995; 117: 2732-2737.
47. Khalaf N, Govardhan C, Lalonde JJ et al. Cross-linked enzyme crystals as highly active catalysts in organic solvents. J Am Chem Soc 1996; 118:5494-5495.
48. Tuengler P, Pfleiderer G. Enhanced heat, alkaline and tryptic stability of acetamidinated pig heart lactate dehydrogenase. Biochim Biophys Acta 1977; 484: 1-8.
49. Torchilin VP, Maksimenko AV, Smirnov VN et al. The principles of enzyme stabilization IV. Modification of 'key' functional groups in the tertiary structure of proteins. Biochim Biophys Acta 1979; 567:1-11.
50. Cupo P, El-Deiry WS, Whitney PL et al. Stability of acetylated and superguanidinated chymotrypsinogens. Arch Biochem Biophys 1982; 216:600-604.
51. Cupo P, El-Deiry WS, Whitney PL et al. Stabilization of proteins by guanidination. J Biol Chem 1980; 255:10828-10833.
52. Mozhaev VV, ˇSikˇsnis VA, Melik-Nubarov NS et al. Protein stabilization via hydrophilization. Covalent modification of trypsin and alpha-chymotrypsin. Eur J Biochem 1988; 173:147-154.
53. Melik-Nubarov NS, Mozhaev VV, Siksnis S et al. Protein stabilization via hydrophilization: stabilization of alpha-chymotrypsin by reductive alkylation with glyoxylic acid. Biotechnol Lett 1987; 9:725-730.
54. Mozhaev VV, Melik-Nubarov NS, Levitsky VY et al. High stability to irreversible inactivation at elevated temperatures of enzymes covalently modified by hydrophilic reagents: alpha-chymotrypsin. Biotechnol Bioeng 1992; 40: 650-662.
55. Siksnis V, Mozhaev VV, Galkantaite N et al. Unusual (zig-zag) temperature dependence of the rate constant for irreversible thermoinactivation of hydrophilized enzymes. Collect Czech Chem Commun 1990; 55:1366-1371.
56. Murphy A, Ó Fágáin C. Stability charactersitics of chemically modified soluble trypsin. J Biotechnol 1996; 49:163-171.
57. Murphy A, Ó Fágáin C. Dipeptide synthesis using acetylated trypsin derivative. Biotechnol Techniques 1997; 11:13-16.
58. Tsuji RF. Enhanced enzymatic activity and stability of trypsin by reductive alkylation in solid phase. Biotechnol Bioeng 1990; 36:1002-1005.
59. Smith RAG, Dewdney JM, Fears R et al. Chemical derivatization of therapeutic proteins. Trends Biotechnol 1993; 11:397-403.
60. Inada Y, Takahashi K, Yoshimoto T et al. Application of polyethylene glycol-modified enzymes in biotechnological processes: organic solvent-soluble enzymes. Trends Biotech 1986; 4:190-194.
61. Inada Y, Yoshimoto T, Matsushima A et al. Engineering physicochemical and biological properties of proteins by chemical modification. Trends Biotechnol 1986; 4:68-73.
62. Inada Y, Furukawa M, Sasaki H et al. Biomedical and biotechnological applications of PEG- and PM-modified proteins. Trends Biotechnol 1995; 13:86-91.

63. Beauchamp CO, Gonias Sl, Menapace DP et al. A new procedure for the synthesis of polyethylene glycol-protein adducts; effects on function, receptor recognition, and clearance of superoxide dismutase, lactoferrin and alpha-2-macroglobulin. Anal Biochem 1983; 131: 25-33.
64. Francis GE, Delgado C, Fisher D. PEG-modified proteins. In: Ahern TJ, Manning MC, eds. Stability of Protein Pharmaceuticals. Part B: In Vivo Pathways of Degradation and Strategies for Protein Stabilization. NY: Plenum Press, Inc., 1992:235-263.
65. Gaertner H, Puigserver AJ. Increased activity and stability of poly(ethylene glycol)-modified trypsin. Enzyme Microb Technol 1992; 14:150-155.
66. Yang Z, Domach M, Auger R et al. Polyethylene glycol-induced stabilization of subtilisin. Enzyme Microb Technol 1996; 18:82-89.
67. Yang Z, Williams D, Russell AJ. Synthesis of protein-containing polymers in organic solvents. Biotechnol Bioeng 1995; 45:10-17.
68. Christensen TB, Vegarud G, Birkeland AJ. Stabilization of enzymes by glycosylation. Process Biochem 1976; 11:25-26.
69. Marshall JJ, Rabinowitz ML. Enzyme stabilization by covalent attachment of carbohydrate. Arch Biochem Biophys 1975; 167:777-779.
70. Srivastava RAK. Studies on stabilization of amylase by covalent coupling to soluble polysaccharides. Enzyme Microb Technol 1991; 13:164-170.
71. Germain P, Slagmolen T, Crichton RR. Relation between stabilization and rigidification of the three-dimensional structure of an enzyme. Biotechnol Bioeng 1989; 33:563-569.
72. Lenders J-P, Crichton RR. Thermal stabilization of amylolytic enzymes by covalent coupling to soluble polysaccharides. Biotechnol Bioeng 1984; 26: 1343-1351.
73. Yamagata Y, Arakawa K, Yamaguchi M et al. Functinal changes of dextran-modified alkaline proteinase from alkalophilic *Bacillus* sp. Enzyme Microb Technol 1994; 16:99-103.
74. Longo MA, Meynial I, Combes D. Chemical and enzymatic glycosylation of enzymes. Ann NY Acad Sci 1995; 750: 125-129.
75. Tor R, Dror Y, Freeman A. Enzyme stabilization by bilayer 'encagement'. Enzyme Microb Technol 1989; 11:306-312.
76. Shami EY, Rothstein A, Ramjeesingh M. Stabilization of biologically active proteins. Trends Biotechnol 1989; 7:186-190.
77. Shami EY, Ramjeesingh M, Rothstein A et al. Stabilization of enzymes by their specific antibodies. In: van den Tweel WJJ, Harder A, Buitelaar RM, eds. Stability and Stabilization of Enzymes. Amsterdam: Elsevier, 1993:153-163.
78. Shami EY, Ramjeesingh M, Rothstein A et al. Stabilization of enzymes by their specific antibodies. Enz Microb Technol 1991; 13:424-428.
79. Tijssen P. Practice and Theory of Enzyme Immunoassays. In: Burdon RH, van Knippenberg PH, eds. Laboratory Techniques in Biochemistry and Molecular Biology 15. Amsterdam: Elsevier, 1985.
80. Oliver JM, Berlin RD, Davis BH. Use of horseradish peroxidase and fluorescent dextrans to study fluid pinocytosis in leukocytes. In: Sabato GD, Langone JJ, Van Vunakis H, eds. Meths Enzymol vol 108 NY: Academic Press, 1984:336-347.
81. Renz M, Kurz C. A colorimetric method for DNA hybridization. Nucl Acids Res 1984; 12:3435-3444.
82. Childs RE, Bardsley WG. The steady-state kinetics of peroxidase with 2,2'-azino-di-(3-ethyl-benzthiazoline-6-sulphonic acid) as chromogen. Biochem J 1975; 145:93-103.
83. Bos ES, van der Doelen AA, van Rooy N et al. 3,3',5,5'-tetramethylbenzidine as an Ames test-negative chromogen for horseradish peroxidase in enzyme immunoassay. J Immunoassay 1981; 2:187-204.

84. Whitehead TP, Thorpe GHG, Carter TJN et al. Enhanced luminescence procedure for sensitive determination of peroxidase-labelled conjugates in immunoassay. Nature 1983; 305:158-159.
85. Thorpe GHG, Kricka LJ, Gillespie E et al. Enhancement of the horseradish peroxidase-catalyzed chemiluminescent oxidation of cyclic diacyl hydrazides by 6-hydroxybenzothiazoles. Anal Biochem 1985; 145:96-100.
86. Sanchez PD, Blanco PT, Fernandez-Alvarez JM et al. Flow-injection analysis of hydrogen peroxide using a horseradish peroxidase-modified electrode detection system. Electroanalysis 1990; 2: 303-308.
87. Blum LJ. Chemiluminescence flow injection analysis of glucose in drinks with a bienzyme fiber optic biosensor. Enz Microb Technol 1993: 15:407-411.
88. Klibanov AM, Berman Z, Alberti BN. Preparative hydroxylation of aromatic compounds catalyzed by peroxidase. J Amer Chem Soc 1981; 103:6263-6234.
89. Schwartz RD, Hutchinson DB. Microbial and enzymatic production of 4,4'-dihydroxybiphenyl via phenol coupling. Enz Microb Technol 1981; 3:361-364.
90. Dordick JS. Enzymatic and chemoenzymatic approaches to polymer synthesis. Trends Biotechnol. 1992; 10:287-293.
91. Dordick JS. Nonaqueous enzymology. In: Blanch HM, Clark DS, eds. Applied Biocatalysis 1. NY: Marcel Dekker Inc, 1991:1-51.
92. Kazandjian RZ, Dordick JS, Klibanov AM. Enzymatic analyses in organic solvents. Biotechnol Bioeng 1986; 28: 417-421.
93. Takahashi K, Nishimura H, Yoshimoto T et al. A chemical modification to make horseradish peroxidase soluble and active in benzene. Biochem Biophys Res Commun 1984; 121:261-265.
94. Miland E, Smyth MR, Ó Fágáin C. Phenol removal by modified peroxidases. J Chem Tech Biotechnol 1996; 67:227-236.
95. Welinder KG. Amino acid sequence studies of horseradish peroxidase. Amino and carboxyl termini, cyanogen bromide and tryptic fragments, the complete sequence and some structural characteristics of horseradish peroxidase. Eur J Biochem 1979; 96:483-502.
96. Kurosaka A, Yana A, Itoh N et al. The structure of a neural specific carbohydrate epitope of HRP recognized by anti-HRP serum. J Biol Chem 1991; 266:4168-4172.
97. Dunford HB. Horseradish peroxidase: structure and kinetic properties. In: Everse J, Everse KE, Grisham MB, eds. Peroxidases in Chemistry and Biology, vol 2. Boca Raton: CRC Press, 1991:1-24.
98. Gilfoyle DJ, Rodriguez-Lopez JN, Smith AT. Probing the aromatic-donor-binding site of horseradish peroxidase using site-directed mutagenesis and the suicide substrate phenylhydrazine. Eur J Biochem 1996; 236:714-722.
99. Rodriguez-Lopez JN, Smith AT, Thorneley RNF. Role of arginine 38 in horseradish peroxidase. A critical residue for substrate binding and catalysis. J Biol Chem 1996; 271:4023-4030.
100. Chiswell DJ, Ortlepp SA. New synthetic DNA encoding horseradish peroxidase C isoenzyme. European Patent EP 299682: 1989.
101. Smith AT, Santama N, Dacey S et al. Expression of a synthetic gene for horseradish peroxidase C in *Escherichia coli* and folding and activation of the recombinant enzyme with Ca^{2+} and heme. J Biol Chem 1990; 265:13335-13343.
102. Welinder KG. Plant peroxidases. their primary, secondary and tertiary structures, and relation to cytochrome *c* peroxidase. Eur J Biochem 1985; 151:497-504.
103. Ryan O, Smyth MR, Ó Fágáin C. Thermostabilized chemical derivatives of horseradish peroxidase. Enzyme Microb Technol 1994; 16:501-505.
104. Miland E, Smyth MR, Ó Fágáin C. Modification of horseradish peroxidase with bifunctional N-hydroxysuccinimide esters: effects on molecular stability. Enzyme Microb Technol 1996; 19:242-249.

105. Miland E, Smyth MR, Ó Fágáin C. Increased thermal and solvent tolerance of acetylated horseradish peroxidase. Enzyme Microb Technol 1996; 19:63-67.
106. Ugarova NN, Rozhkova GD, Berezin IV. Chemical modification of the epsilon-amino groups of lysine residues in horseradish peroxidase and its effect on the catalytic properties and thermostability of the enzyme. Biochim Biophys Acta 1979; 570:31-42.
107. Vazquez-Duhalt R, Fedorak PM, Westlake DWS. Role of enzyme hydrophobicity in biocatalysis in organic solvents. Enzyme Microb Technol 1992; 14: 837-841.
108. Bonnaffe D, Therisod M, Seris JL et al. Chemical deglycosylation of horseradish peroxidase and surglycosylation using a new glycosylating reagent: effects on catalytic activity and stability. In: van den Tweel WJJ, Harder A, Buitelaar RM, eds. Stability and Stabilization of Enzymes. Amsterdam: Elsevier, 1993:223-227.
109. Egorov AM, Gazaryan IG, Kim BB et al. Horseradish peroxidase isoenzyme C. A comparative study of native and recombinant enzyme produced by *E. coli* transformants. Ann NY Acad Sci 1994; 721: 73-81.
110. Chang BS, Park KH, Lund DB. Thermal inactivation kinetics of horseradish peroxidase. J Food Sci 1988; 53:920-923.
111. Hendrickx M, Saraiva J, Lyssens J et al. The influence of water activity on thermal stability of horseradish peroxidase. Int J Food Sci, Technol 1992; 27:33-40.
112. Kirkwood TBL. Design and analysis of accelerated degradation tests for the stability of biological standards III. Principles of design. J Biol Stand 1984; 12:215-224.
113. Pace CN. Measuring and increasing protein stability. Trends Biotechnol 1990; 8:93-98.
114. Pappa HS, Cass AE. A step towards understanding the folding mechanism of HRP. Eur J Biochem 1993; 212:227-235.
115. Haschke RH, Friedhoff JM. Calcium related properties of horseradish peroxidase. Biochem Biophys Res Commun 1978; 80:1039-1042.
116. Ryu K, Dordick JS. How do organic solvents affect peroxidase structure and function? Biochemistry 1992; 31: 2588-2598.
117. Gorman LA, Dordick JS. Organic solvents strip water off enzymes. Biotechnol Bioeng 1992; 39:392-397.
118. Hiner ANP, Hernández-Ruíz J, García-Cánovas F et al. A comparative study of the inactivation of wild-type, recombinant and two mutant horseradish peroxidase isoenzymes C by hydrogen peroxide and m-chloroperoxybenzoic acid. Eur J Biochem 1995; 234:506-512.
119. Yeoman JC, Hager LP. Isolation of a heme crevice peptide from affinity-labelled horseradish peroxidase. Biochem Biophys Res Commun 1980; 97: 1233-1240.
120. Khmelnitsky YLK, Mozhaev VV, Bolova AB et al. Denaturation capacity: new quantitative criterion for selection of organic solvents as reaction media in biocatalysis. Eur J Biochem 1991; 198: 31-41.
121. Martinez P, Arnold FH. Surface charge substitutions increase the stability of alpha-lytic protease in organic solvents. J Am Chem Soc 1991; 113:6336-6337.
122. Zaks A, Klibanov AM. The effect of water on enzyme action in organic media. J Biol Chem 1988; 263:8017-8021.
123. Batra R, Gupta MN. Enhancement of enzyme activity in aqueous-organic solvent mixtures. Biotechnol Lett 1994: 10:1059-1064.
124. Yoshitake S, Imagawa M, Ishikawa E et al. Mild and efficient conjugation of rabbit Fab' and horseradish peroxidase using a maleimide compound and its use for enzyme immunoassay. J Biochem (Tokyo) 1982; 92:1413-1424.
125. Nilsson P, Bergquist NR, Grundy MS. A technique for preparing defined conjugates of horseradish peroxidase and immunoglobulin. J Immunol Meth 1981; 41:81-93.

126. Arseguel D, Lattes A, Baboulene M. Synthesis by enzymatic catalysis VI - an investigation of chemical modification of HRP by Fourier transform infra-red vibrational spectroscopy. Biocatalysis 1990; 3:217-225, 227-233.
127. Urrutigoity M, Souppe J. Biocatalysis in organic solvents with a polymer-bound horseradish peroxidase. Biocatalysis 1989; 2:145-149.
128. Wirth P, Souppe J, Tritsch D et al. Chemical modification of horseradish peroxidase with ethanal-methoxypolyethylene glycol: solubility in organic solvents, activity and properties. Bioorganic Chem 1991; 19:133-142.
129. Souppe J, Urrutigoity M, Levesque G. Application of the reaction of dithioesters with epsilon-amino groups in lysine to the chemical modification of proteins. Biochim Biophys Acta 1988; 957:254-257.
130. Roig MG, Kennedy JF. Perspectives for chemical modifications of enzymes. CRC Crit Rev Biotechnol 1992; 12:391-412.

CHAPTER 7

Immobilization

Introduction

The immobilization of enzymes and proteins onto insoluble materials forms the basis of many biotechnological processes and analytical devices. It is in its own right an extremely active field of applied study, so only the functional stability aspects of enzyme and protein immobilization will be considered here. There are four basic immobilization processes, namely: physical adsorption, covalent binding, copolymerization and microencapsulation.[1] (See Fig. 7.1.) Each of these methods can (or may not!) bring about increased stability of a protein or enzyme, but some are better in this regard (as will be discussed). It is important to remember, however, that immobilization is often performed for other reasons (such as ease of enzyme-product separation, or enzyme recyclability) and not primarily to stabilize the protein or enzyme of interest. Thus, any improved stability characteristics may be merely a useful bonus arising from an immobilization protocol dictated by other considerations. Readers with a greater interest in immobilization should consult references 1-4. A host of other monographs, reviews (including refs. 5-7) and research papers are available.

Among the novel properties of immobilized enzymes initially noted was an increase in stability or persistence of functional activity vis-à-vis their soluble counterparts. For instance, Chang observed a greatly lengthened persistence of activity for catalase microencapsulated with hemoglobin as a carrier protein. Soluble catalase under the same storage conditions retained much less of the original activity. A glutaraldehyde-crosslinked microencapsulated preparation of catalase proved to be even more stable than the plain microencapsulated derivative.[8] Johnson has reviewed the evidence for some reportedly stabilized immobilized enzymes. In some cases the reduction of autodigestion was confused with thermostabilization.[9] Norde has studied protein adsorption to solid surfaces. Accumulation of charges in the region of protein binding to the support tends to be relieved by a transfer of low molecular weight ions to the adsorbed layer from the bulk solution. This favorable electrical effect is opposed by an unfavorable chemical effect. Where the solid phase is hydrophobic, adsorption of the much larger protein molecules will release many more water molecules from the solid surface, favoring protein immobilization by an increase in the entropy of water molecules. Norde argues that proteins can undergo structural rearrangements during adsorption, citing evidence that some adsorbed proteins show notably-decreased secondary structure. He also classifies protein molecules as 'hard' or 'soft'. Hard proteins undergo little structural change on adsorption to surfaces; thus, their binding is dictated by electrostatic

Stabilizing Protein Function, by Ciarán Ó Fágáin.
© Springer - Verlag and Landes Bioscience 1997.

attractions and by dehydration of the solid surface. Accordingly, hard proteins can bind to hydrophobic surfaces and can adsorb also onto hydrophilic surfaces if there is an electrostatic attraction. Soft proteins, however, have a stronger affinity for surfaces: structural rearrangements (perhaps quite extensive) occurring within the protein molecule upon adsorption provide a strong driving force for the adsorption process. Because of this, soft proteins are capable of adsorbing even to a hydrophilic, electrostatically-repelling surface. As an illustration, alpha-lactalbumin in the presence of calcium ions bears a net negative charge and adsorbs poorly to negatively-charged SiO_2 particles. Removal of Ca^{2+} ions destabilizes the alpha-lactalbumin molecules, facilitating structural changes upon adsorption. Under these conditions, a much greater binding of alpha-lactalbumin to SiO_2 takes place.[10]

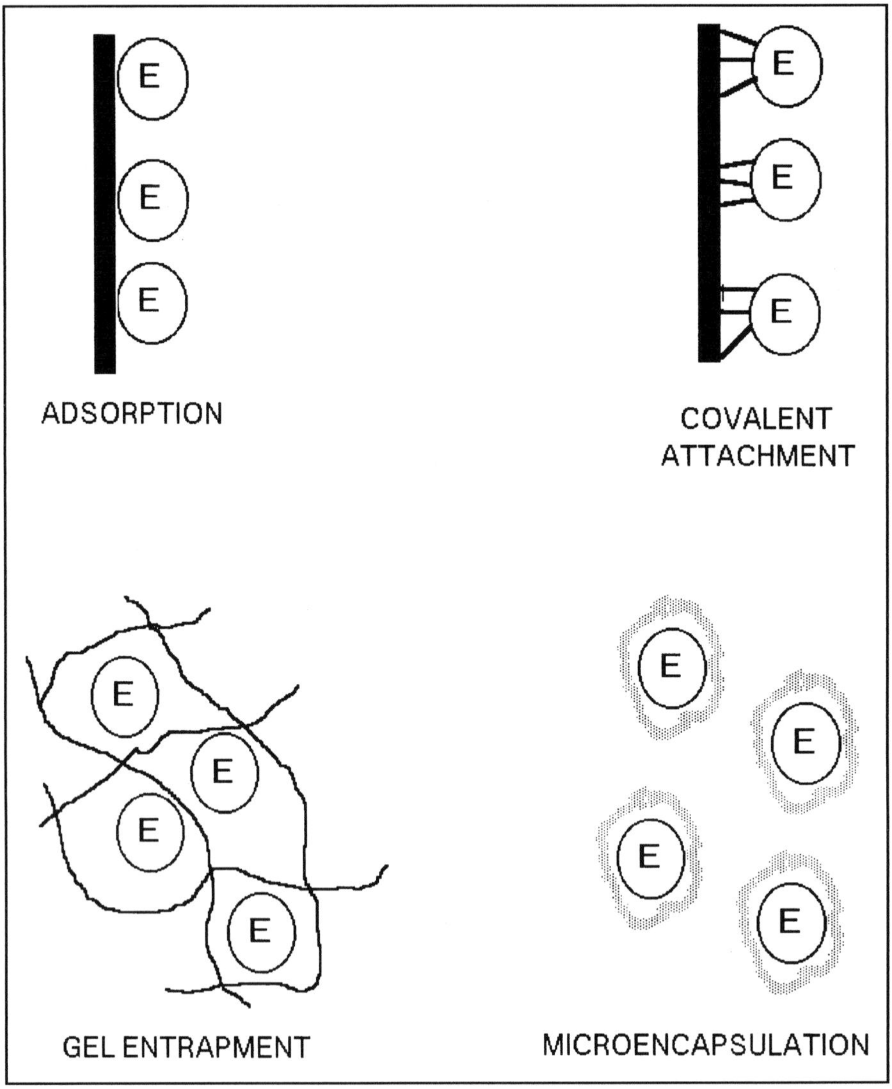

Fig. 7.1. Some of the possible methods for immobilization of proteins.

Investigation of 'Immobilization Thermostability'

Some studies have focused specifically on the enhanced (thermo)stability of immobilized enzyme derivatives.[11,12] Martinek et al stabilized chymotrypsin and trypsin by gel immobilization[11] (copolymerization) based on work by Gabel.[13] The enzymes were initially acylated with acryloyl chloride or alkylated with acrolein. Derivatized enzymes were then copolymerized with sodium methacrylate or with acrylamide. The enzymes immobilized in this way showed dramatically increased thermostabilities compared with their soluble, native forms. Comparison of apparent first order-rate constants indicated that the thermostabilities of the immobilized enzyme derivatives had increased by several hundred-fold over the range 60-100°C. Each enzyme was attached to the polymer support at multiple locations and the greater the number of attachment points (i.e. the greater the degree of modification in the first chemical treatment stage), the greater the degree of stabilization achieved. The stability gains were attributed to rigidification of the protein molecule. Greater rigidity prevents the enzyme molecule from unfolding and there is much less deterioration of activity in the face of denaturing influences such as heat.

The same group of researchers demonstrated that simple entrapment of chymotrypsin and trypsin within a polymethacrylate gel can also lead to significant thermostabilization.[12] Chymotrypsin entrapped within a 50% w/w polymethacrylate gel proved to be 100,000 times more stable than the native, based on comparison of the apparent first-order thermoinactivation rate constants. The inactivation rate constant for trypsin immobilized in polymethacrylate gel drops almost to zero as the gel concentration rises above 25% w/w. This stabilization was ascribed to the formation of multiple noncovalent linkages (electrostatic or hydrogen bonds) between the enzyme and the small pore-size polyelectrolyte gel. In contrast, polyacrylamide gel is electrically neutral and stability is independent of gel concentration up to 50% w/w. This proved that physical or mechanical immobilization can be just as effective a stabilizing strategy as chemical immobilization. These copolymerizations prevented the reversible denaturation of enzyme activity but this immobilization methodology can also stabilize proteins against irreversible inactivation. Mozhaev and colleagues have compared copolymerization with other immobilization techniques.[14] Immobilized trypsin had half life exceeding 10 hours at 75°C in the absence of substrate and seemed even more stable when incubated with substrate. Soluble trypsin inactivated almost immediately at 75°C. Both the temperature of maximum activity and the rate constant of irreversible thermoinactivation of trypsin depended strongly on the number of bonds formed between the enzyme and the solid phase. This implies that the immobilization process has strengthened trypsin against activity loss due to some (reversible) perturbation of the active site and also against gross (essentially irreversible) molecular unfolding. Mozhaev and coworkers have succinctly reviewed enzyme stabilization by copolymerization techniques. One can incorporate enzymes into charged gels or into hydrophobic polymers (which are compatible with organic solvents). Enzyme incorporation into methacrylate gels may be regulated by salts.[15]

Some Specific Examples

The examples outlined below highlight aspects of enzyme stabilization by immobilization.

Soluble and insoluble polymers of N-vinyl pyrrolidone with glycidyl acrylates and glycidyl methacrylates have been used to stabilize lipase and amylase.[16] These polymers have many reactive epoxide groups. A mixed-polymer solid phase was synthesized prior to coupling of lipase at pH 9.1 and of amylase over a range of pH values. Both enzymes, but especially the lipase, were much more stable at 50°C while the pH optimum of each increased by about 0.5 units. Both enzymes showed decreased activity following immobilization (lipase as low as 0.3%) but the remaining activity was greatly stabilized.

Guisán and colleagues have examined the factors influencing stabilization of trypsin by covalent attachment to agarose particles.[17,18] They enumerated four main factors affecting the number of links between the trypsin molecule's amine groups and the aldehyde functions of the agarose solid phase; these were: number of aldehyde groups per unit area of activated gel, pH of the medium, reaction time (for coupling of enzyme to activated solid phase) and temperature. The Guisán group prepared a number of trypsin-agarose derivatives using different combinations of these four variables. Inclusion of a competitive inhibitor in the reaction mixture led to complete recovery of starting activity but also to derivatives with widely-varying stability characteristics. Enzyme molecules immobilized by a single covalent bond were no more stable than their soluble counterparts. (Effectively, of course, such a molecule is merely 'tethered' to the support; there is nothing to prevent its denaturation or inactivation, with the possible exception of steric hindrance). The most stable derivative proved to be 5,000 times more stable than soluble trypsin. Seven of its lysine residues had reacted with the solid phase. (The optimal density of aldehyde groups on the surface of the solid phase was 20 per 100 square nanometers.) Alkaline pH (up to pH 10) favored stabilization. This is probably due to the increase in the number of unprotonated amino groups on the protein surface with increasing pH. It is these uncharged amino functions which react with the support's aldehyde groups. Optimum conditions for stabilization were 72 hours at 25-35°C. This is quite a long period and it implies that formation of the optimal stabilizing interactions between the protein and the solid phase proceed slowly, even though individual amino group-aldehyde reactions occur rapidly under these same conditions. This suggests in turn that some sort of chemical 'self stabilization' occurs with time, echoing that observed in experiments involving a cocktail of bifunctional crosslinkers of varying length (see refs. 19 and 20).

Guisán and colleagues argue that multipoint covalent immobilization to a pre-existing aldehyde-rich support in this manner represents a workable immobilization/stabilization system. Porous particles containing internal spaces have a much greater surface area, and afford many more locations (both surface and internal) for protein attachment than do non-porous solids. Accordingly, it should be relatively easy to attain a high number of links between the target protein and the support, thereby rigidifying and stabilizing the enzyme. There is a wide variety of porous supports with differing mechanical properties, including 'biological' polymers (such as agarose and cellulose) and inorganic substances (e.g. alumina, porous glass, silica and titania). The target protein's lysines seem to be the residues of choice for immobilization: lysines occur frequently in proteins and are active nucleophiles when unprotonated, i.e., at alkaline pH. Their charged/polar nature means that they occur on protein surfaces and they do not need to be activated before the immobilization reaction. Aldehydes are particularly suitable as active groups on the solid particles: they are a good target for nucleophilic attack and lack sterically-hindering substituents. They are stable and one can incorporate many reactive groups onto the support particle. Activation is achieved by epoxypropanol etherification of the agarose gel to yield a glyceryl intermediate. This intermediate is then oxidized by periodate to the desired aldehyde function. These chemical reactions permit a very dense placing of active aldehyde groups on the surface of the agarose gel particle. Indeed, it is claimed that more than 30 active aldehyde groups are present on the area of particle surface occupied by a single trypsin molecule.[21] It can also be convenient and helpful to immobilize/stabilize one's target protein in this manner before undertaking chemical modification.[21,22] Enzymes stabilized by this strategy include trypsin, chymotrypsin, penicillin acylase (see below), beta-galactosidase and lipase.[21]

An Immobilized Biocatalyst: Penicillin Amidase (PGA)

Some immobilized enzymes form the basis of routine and important industrial processes. Examples include glucose isomerase (to manufacture high-fructose syrups for the food industry: see chapters 4 and 8) and penicillin amidase (also known as penicillin acylase), used to produce semisynthetic antibiotics. The fundamental reason for immobilization of these two enzymes, however, was to permit their re-use, with concomitant reduction in process costs, rather than to stabilize them.[23] For industrial use, candidate solid phase particles should be of a suitable size, have good mechanical properties and have a high loading capacity for the desired enzyme or protein. They should be able to bind the target enzyme efficiently (preferably in a single step reaction) and, once this has been accomplished, should provide an effective stabilizing support for that enzyme. Typical target performance parameters would be a total product yield of many hundreds of kilograms for every kilogram of immobilized enzyme. Also, the biocatalyst should have a storage half life of at least 6 months.[24]

Beta-lactam antibiotics, such as naturally-occurring penicillins and cephalosporins, are important bacteristatic drugs which prevent the final step in bacterial cell wall synthesis, leading to cell lysis and death. A number of man-made derivatives with improved properties have been prepared. Among these 'improvements', one may cite increased acid resistance (allowing oral administration, since some modified penicillins can tolerate stomach acid) or effectiveness against a broader range of bacteria (some derivatives can act against Gram-negative bacteria, unlike many naturally-occurring forms). A critical feature of these antibiotic molecules is their beta-lactam ring structure (Fig. 7.2): its rupture inactivates the antibiotic. Penicillin amidase permits specific manipulation of natural penicillins and cephalosporins under mild conditions, preserving the beta-lactam ring and resulting in high yields of the desired semisynthetic antibiotic. The enzyme is doubly useful. At pH 7.5-8.5, it cleaves the naturally-occurring side chain or R-group from the penicillin molecule, yielding 6-aminopenicillanic acid[25] (or 6-APA; see Fig. 7.2.). 6-APA has no antibiotic activity itself but acts as an intermediate for the production of semisynthetic derivatives. Under different conditions (pH 4.0-7.0), penicillin amidase adds a new, desired R-group to the 6-APA core, resulting in a new drug with altered (and often improved) properties.[25] (Penicillin amidase can also make semisynthetic cephalosporins via acylation of 7-amino-deacetoxy-cephalosporanic acid[26] [7-ADCA].) It is important that the enzyme be immobilized to permit efficient separation from the 6-APA (which must be cooled and acidified for crystallization and collection)[24] or from the finished semisynthetic. The enzyme is used in conventional stirred-tank vessels incorporating a filtration system at the bottom.[24]

The penicillin amidase enzyme occurs in a number of different microorganisms but the most familiar example is that from *E. coli*. In the cases studied so far,

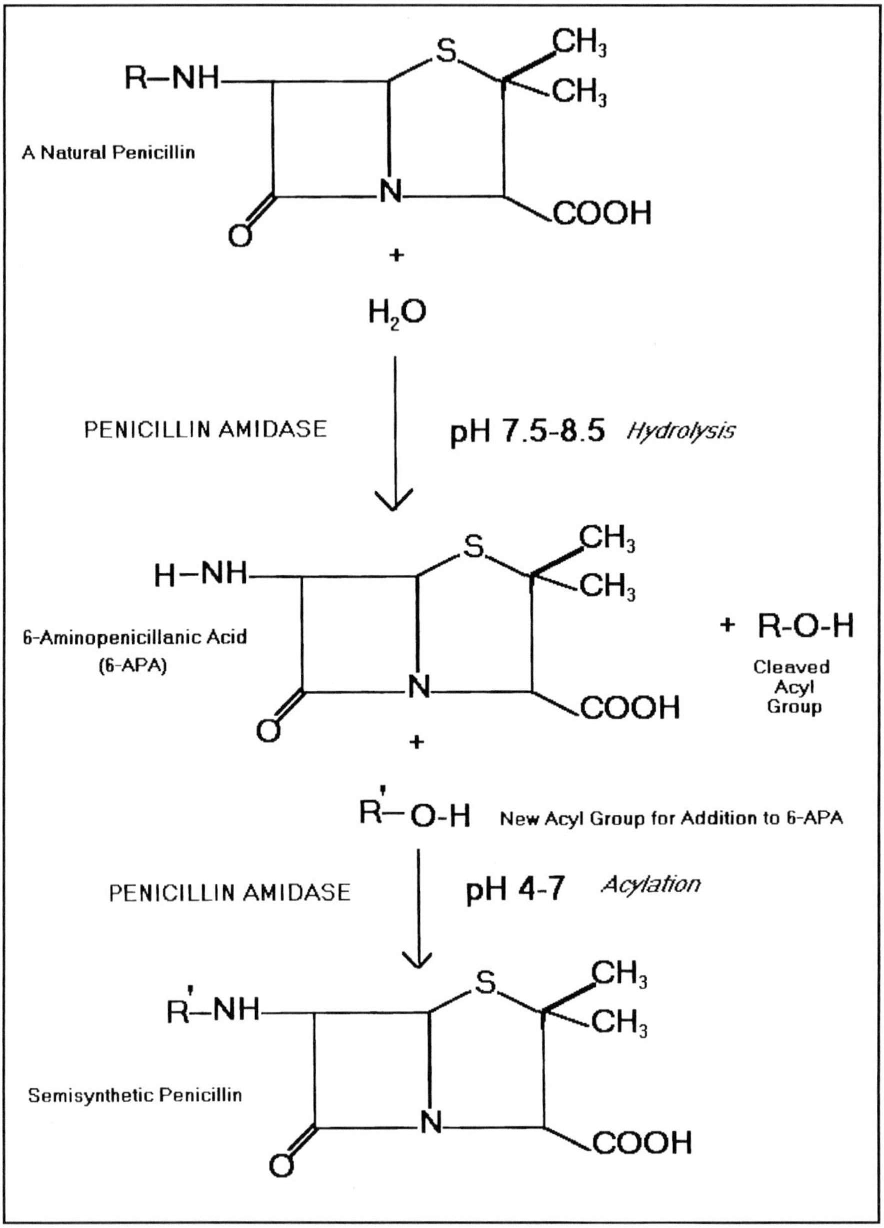

Fig. 7.2. Reactions catalyzed by penicillin amidase: production of 6-aminopenicillanic acid and its acylation to yield a semisynthetic penicillin derivative.

the enzyme has two non-identical subunits. The alpha subunit weighs 23.8 kDa and confers specificity for the penicillin side chain while the 62.2 kDa beta subunit performs the catalytic steps.[24,27,28] Separation of the subunits leads to loss of activity.[28] The alpha subunit contains a hydrophobic patch which binds the 'hydrophobic probe' 8-anilino-2-naphthalene sulfonic acid; it is thought that the beta subunit covers this patch in the native enzyme.[28] Neither subunit has disulfide bonds and the enzyme is unstable at pH values greater than 8. This poses problems for pH maintenance during the hydrolytic reaction.[24] The German company Hoechst AG achieved improved specific activities and yields of penicillin amidase using conventional strain improvement and fermentation techniques, but at the price of disimproved operational stability. (This was possibly due to a stabilizing effect on the crude enzyme preparation of other, non-amidase protein molecules.) The Hoechst scientists selected the commercial acrylic resin Eupergit™, which has a high content (0.8 $mmol.g^{-1}$) of epoxy groups for coupling, as solid phase. This reacts with lysine and tyrosine residues of the target enzyme. Once the Eupergit polymerization reaction had been optimized, the immobilized penicillin amidase displayed very good operational stability characteristics, being good for over 100 recyclings. This performance outstripped that of a vinylacetate polymer onto which epoxide groups were introduced following its polymerization; this latter solid phase had only half the active epoxide groups per gram of the commercial Eupergit. Probably, a lesser number of enzyme-polymer links in the vinyl acetate case accounts for the observed stability differences. Half lives could be stretched further by the use of ammonia to maintain pH rather than sodium hydroxide and by replacement of the stirred tank reactor with a packed bed recycle reactor: the latter arrangement separates the enzyme-substrate reaction from the acid neutralization.[24] Findings such as these show how refinement of process design, in addition to the stabilizing strategies outlined in this volume, can positively influence enzyme or protein stability. Similar improvements in operational stability were recorded for glutaryl amidase following the addition of a buffer, a one-unit shift in processing pH from 7 to 8 and replacement of sodium hydroxide as neutralizing agent by ammonia. Glutaryl amidase is used to produce 7-aminocephalosporanic acid (7-ACA), a key intermediate in the manufacture of semisynthetic cephalosporins, from cephalosporin C.[24] Optimized processes of this sort can be remarkably efficient: it has been calculated that, on a dry weight basis, 1 kg of immobilized penicillin amidase can generate more than 600 kg of 6-APA. This means that enzyme costs are less than 3% of the price of the total amount of 6-APA.[24]

PGA from *E. coli* was successfully immobilized on agarose (aldehyde) gels in a multipoint covalent fashion in similar manner to trypsin. The most stable derivative was approximately 1,400 times more stable than the free enzyme. Important variables in achieving stabilized, multiply-coupled derivatives included the density of aldehyde groups on the surface of the activated support, temperature and the duration of contact between the enzyme and activated support before borohydride reduction of the Schiff's bases formed. Tolerances of heat and of water-methanol mixtures were used to gauge stability. The immobilized PGA derivatives showed multiphasic thermal inactivation kinetics, so true half lives could not be calculated. Instead, comparisons of the various PGA forms were undertaken using pseudo-halftimes (time required to reach 50% residual activity) estimated from the experimental time courses. Immobilization was achieved without any loss of activity. In all cases, the immobilization proved to be irreversible and no subunit desorption was observed even in strongly desorbing conditions using sodium dodecyl sulfate. Contact times in the range 3-24 hours at about 25°C led to the greatest stabilization factors. It seems that the formation of optimally-stabilizing PGA-support contacts is a much slower kinetic process than the initial immobilizing events.[29]

Guisán and colleagues have focussed on the penicillin amidase from *Kluyveromyces citrophila* as well as on that from *E. coli*. The *K. citrophila* enzyme has similar catalytic parameters (V_{max} and K_m) to *E. coli* PGA but much better intrinsic stability characteristics. It is much more resistant to heat, pH and organic cosolvents and is easily immobilized via its free amino groups. In thermal deactivation time course experiments carried out at 58°C over a range of pH values, the *K. citrophila* enzyme consistently performed better. It was 70-, 10- and 2-fold more stable than the *E. coli* PGA at pH 5.0, 8.0 and 10.0, respectively. The greater intrinsic stability of the *K. citrophila* enzyme permitted its successful immobilization at pH 10.0, 15°C without any stabilizing additive.[30] Further work on *K. citrophila* penicillin amidase achieved stabilization factors of more than 1,000 with 90% recovery of starting activity; both subunits were multiply attached to the agarose-aldehyde solid phase.[31]

Guisán's group later showed that a simple, non-crosslinking chemical modification of penicillin amidase *after* immobilization could yield further stability gains. Immobilized-stabilized derivatives of *E. coli* and *K. citrophila* were treated with 0.5% formaldehyde and then reduced with sodium borohydride. The formaldehyde forms Schiff's bases with PGA free amino groups which had not been involved in the previous covalent couplings to solid phase, while borohydride reduces the Schiff's bases to make the modification irreversible. Thus, the chemistry of modification resembles that of the immobilization reaction itself. The formaldehyde-PGA reaction does not form a crosslink; it merely modifies amino groups on the PGA surface. It was important that the PGA was both immobilized *and* stabilized prior to formaldehyde treatment: single point-immobilized (i.e. not stabilized) PGA lost notably more catalytic activity during the formaldehyde modification than did multiply-immobilized (and stabilized) PGA. Borohydride (1 $mg.ml^{-1}$) reduction proved necessary for effective stabilization. Greater stability gains, combined with smaller losses of catalytic activity, took place at neutral to slightly-acidic pH values: a 24-fold stabilization was achieved at pH 5.0 with a very high recovery of initial activity. Reactions with multiple uncharged amino groups will be favored by higher pH values, but the optimal results for formaldehyde modification occurred at pH 5.0. This suggests that reaction of only a very small number of amino groups (10% of the total) leads to the stability gains observed. The 'optimized' conditions for formaldehyde modification/stabilization of *E. coli* PGA were applied to the *K. citrophila* enzyme but here only a 4-fold additional gain in stability characteristics was achieved with a <5% loss of catalytic activity.[22] Thus, the Guisán group has achieved additional stabilization of a commercially-important enzyme by combining two stabilizing methodologies: immobilization and chemical modification. This synergy recalls some similar examples discussed in chapter 6 and mentioned also in chapter 10.

Penicillin G amidase has also been covalently immobilized onto Amberlite XAD7, a commercial poly(methacrylic ester) resin. PGA will adsorb well to Amberlite XAD7 at pH 5-6 but cannot be used at pH 8 due to desorption at this higher pH. Thus, a 'permanent' covalent immobilization is required. The Amberlite was first aminoalkylated with 1,2-diaminoethane, introducing free amino groups which will react with glutaraldehyde. Then the enzyme was bound using glutaraldehyde. This procedure led to the formation of Schiff's bases between the enzyme amino groups and the glutaraldehyde molecules. The stability characteristics of the immobilized PGA improved significantly when these Schiff's bases were reduced using borohydride. This enhanced stability was achieved without alteration to the enzyme's catalytic activity. PGA stored in the presence of its natural substrate, penicillin G, retained more activity than the fraction stored in plain buffer. Immobilized PGA fractions were monitored through repeated reaction cycles. The time required for 50% substrate conversion (penicillin G $\rightarrow$ 6-APA) was used as a measure of operational stability. After 20 reaction cycles, the borohydride-reduced fraction was still 100% active; its unreduced, immobilized-only counterpart retained only 61% activity. (The authors noted that their findings resembled those of ref. 22.) The partial loss of activity by the fraction which was not treated with borohydride was investigated. Neither enzyme activity nor released protein were detected in solution, showing that the activity loss was not due to simple desorption of PGA from the Amberlite support. Column and batch configurations were set up for penicillin G conversion. The penicillin G substrate was not continuously present during batch operations, and this is where the differences were first seen to arise. Experiments showed that the reduced fraction of Amberlite-PGA retained full activity in the presence or absence of substrate while the unreduced fraction lost up to 59% activity in the absence of penicillin G. The decline could be arrested by restoring substrate to the system. Under operating conditions, Amberlite-PGA showed a half life of >500 hours at 37°C.[32]

Other Examples

Creatine kinase plays an important role in clinical diagnosis but the native enzyme is not stable in solution.[33] In spite of this, it has been successfully immobilized and it has potential use as a device for ATP regeneration in enzyme systems.[34] Creatine kinase was immobilized on agarose using CNBr and its thermal stability was determined at 45°C in 0.1 M glycine/NaOH buffer, pH 9.0. The inactivation curve of creatine kinase immobilized at a single point closely resembled that of the soluble enzyme. However, multiply-immobilized creatine kinase retained almost full activity up to 18 hours and was still 50% active after 35 hours. Here the inactivation curve was not first-order.[34] These findings emphasize once again the importance of forming multiple attachments between the target protein and the solid support if one wishes to achieve stabilization.

Gianfreda and Greco pioneered a technique they term 'dynamic immobilization'. This involves the effective immobilization of soluble enzymes on ultrafiltration membranes by the permeate stream in an unstirred bioreactor.[35] The technique has the advantage that no chemical or physical manipulation of the enzyme is required but any stabilization enhancement achieved is marginal. This drawback may be overcome to some extent by the addition to the bioreactor of soluble macromolecules such as dextran, carboxymethyl (CM-) cellulose and water-soluble polyacrylamide. All three polymers and particularly dextran caused some decrease in the measured enzyme activity. Stabilization factors achieved ranged from 3-4-fold (beta-galactose dehydrogenase with dextran) to 22-fold (beta-glucosidase [cellobiase] with CM-cellulose).

One can immobilize enzyme derivatives as well as their native forms. A preparation of cellobiase (beta-glucosidase; hydrolyzes the glucose-beta-1,4-disaccharide known as cellobiose) was stabilized by conjugation to soluble dextran and then immobilized on silica.[36] The cellobiase-dextran conjugate had an optimum temperature 5-10°C higher than that of the native and its activity persisted for twice as long as the free enzyme's. The silica solid phase was activated by amination. Coupling took place on addition of the cellobiase-dextran conjugate. The reductive alkylation coupling procedure took full advantage of the aldehyde groups already introduced onto the enzyme molecule. Glutaraldehyde treatment allowed immobilization of a soluble cellobiase control on the activated support. At 70°C, the immobilized/conjugated cellobiase was more stable than the glutaraldehyde-immobilized unconjugated enzyme. The immobilized/conjugated cellobiase was also much superior in terms of its successive operational stability, i.e., its ability to catalyze cellobiose hydrolysis when introduced into a fresh reaction mixture.

Coulet and colleagues successfully immobilized aspartate aminotransferase on collagen films. The derivatized enzyme retained 100% activity for 5 months on storage at 4°C despite repeated assays at 30°C and associated washings.[37]

The related alanine aminotransferase (ALT) was immobilized by covalent binding to two types of preactivated hydroxylic support (adipic acid dihydrazide agarose, activated by a carbodiimide, binds to carboxyl groups of the protein, and 6-aminohexanoic acid N-hydroxysuccinimide ester Sepharose™, which binds to protein amino groups) and by entrapment in calcium alginate gels. Half life at 50°C of the alginate derivative was 2.75 times greater than that of free ALT. However, the covalently-bound ALT fractions displayed greater stabilization factors: the adipic acid dihyrazide and hexanoic acid derivatives were 7.8 and 5.7 times more tolerant of 50°C incubation than native ALT. The pH optimum (7.5) was identical for all fractions and immobilized fractions performed satisfactorily in batchwise synthesis of L-alanine.[38] Moreno also immobilized an ALT fraction previously modified with dimethylsuberimidate (DMS) onto adipic acid dihydrazide-agrose. DMS reacts with ALT amino groups and the subsequent immobilization took place via ALT carboxyls. The modified-immobilized ALT was more stable than the immobilized-only fraction under a variety of conditions (50°C, 30% dioxane, 0-5 M urea).[39] The additional stabilization of Moreno's chemically-modified ALT preparation by the subsequent immobilization procedure echoes the findings of Lenders et al (ref. 36 and above), whose immobilized, dextran-conjugated cellobiase preparation was much more stable than the corresponding unconjugated, immobilized enzyme. These reports show that stabilizing strategies may be combined to achieve successive increments in stability. (This point is mentioned in chapter 10 also.)

Note also that St. Clair and Navia have prepared entirely novel 'immobilized' enzyme derivatives. These crosslinked enzyme crystals (CLECs™) are greatly stabilized and, due to their crystal nature, take the form of macroscopic particles. Thus, they can be regarded as immobilized enzymes.[40-42] However, the main features of CLECs™ have been described in chapter 6 of this volume rather than in this one, due to the crosslinking chemistry involved and the absence of 'external' macroscopic solid particles from the methodology.

In spite of the above successful examples of enzyme stabilization by immobilization, one must always remember that immobilization reactions need not necessarily result in increased stability. Bickerstaff has warned of this danger.[5] It can be important also to specify what type of stability/stabilization is in question. For instance, immobilization of glucose phosphate isomerase yielded no notable difference in thermal stability but did result in greater resistance to urea denaturation.[43]

Enzyme Deactivation and Reactivation

Immobilized enzymes have contributed enormously to the study of enzyme deactivation and reactivation (see ref. 44 for a review). Refs. 5 and 6 also contain summaries of deactivation/reactivation studies.

In general, a protein may be reactivated if its inactivation resulted from changes in conformation, aggregation state or chemical structure (in practice, this latter concerns modified -S-S- and -SH groups only; see also chapter 2). One can devise suitable treatments to reverse these molecular changes and so reactivate the enzyme. (It is conformational denaturation that is most important in the case of immobilized enzymes: aggregation, autolysis and adsorption are less likely to affect immobilized proteins.) Kinetic reasons may prevent an inactivated enzyme from re-adopting its native conformation on transfer to non-denaturing conditions. Using denaturing solutes, one can unfold an immobilized (or 'tethered') protein to a random or near-random coil. Under appropriate, changed conditions, the unfolded protein may spontaneously refold to its correct functional (active) conformation. Mozhaev and colleagues describe some examples in detail in ref. 44, including the repeated deactivation and reactivation of alpha-chymotrypsin entrapped in polyacrylamide gel. In studies of this kind, one must ensure that the unfolded protein does not react non-specifically with the solid support.[6]

A variant of this technique involves the isolation of individual denatured protein molecules within reverse micelles. The protein can refold to its native conformation inside the micelle without making contact with, and aggregating with, other denatured molecules. The micelles in question consist of aqueous phase droplets stabilized by a surfactant and suspended in 2,2,4-trimethylpentane (formerly known as iso-octane).[45,46]

A further significant observation was the reactivation of enzyme activity following removal of denaturing agents, as reported in ref. 47 for rabbit muscle lactate dehydrogenase (LDH), a tetrameric enzyme. Individual immobilized subunits were thought to retain activity which was enhanced by reassociation with other subunits, but this conclusion was later questioned. Chan and Mosbach demonstrated that individual immobilized rabbit muscle LDH subunits were inactive but could reassociate with added soluble subunits to yield an active derivative.[48]

Enzyme immobilization is a viable methodology for protein stabilization provided one takes certain precautions. One must protect labile or essential groups on the target protein with suitable reagents or ligands. Reaction conditions used should ensure the formation of multiple bonds between the enzyme and the solid phase. The scientist must be ready to accept some alteration in protein properties following immobilization. Typical characteristics affected include activity, substrate binding and pH optimum.[7]

As with other aspects of protein science, recombinant DNA techniques have made an impact on immobilization options and strategies. Alpha-glucosidase can cleave the glycosidic link within the disaccharide maltose to yield glucose. Stempfer and colleagues have described[49] a fusion protein of yeast alpha-glucosidase embodying an $(Arg)_6$ fusion peptide extension at its C-terminus. This polycationic extension permits specific and orientated adsorption of the fusion protein from crude extracts onto polyanionic adsorbents under low-salt conditions (50 mM NaCl or less). The wild type does not adsorb under these conditions, since it lacks the cationic peptide extension. The immobilized fusion protein closely resembled (but did not exactly match) its soluble counterpart in terms of denaturation by urea (resistant up to 1.5 M concentration) and heat inactivation. The pH optimum of the immobilized form underwent a slight alkaline shift. Operational stability of the immobilized form during continuous substrate conversion was notably greater than that of its unattached counterpart. Alpha-glucosidase immobilized on a commercial anionic matrix and operated in a continuous fixed bed reactor showed no notable decrease in substrate conversion throughout 10 days at 30°C; the soluble enzyme is completely inactivated under these conditions. Immobilization may give rise to this improved functional stability either by preventing alpha-glucosidase precipitation at 30°C or by preventing proteolytic inactivation. The specific activity, K_m and activation energy of the enzymatic reaction were unaffected by the immobilization. The fusion protein can be detached by high salt (>0.3 M NaCl). This sort of fusion protein methodology should be widely applicable in the production of immobilized biocatalysts. One must remember, of course, that enzyme leakage remains a hazard for non-covalent immobilization techniques.[49] The C-terminal fusion peptide also facilitates refolding of the immobilized alpha-glucosidase fusion protein without the danger of aggregation. Under optimum conditions, high yields of refolded, immobilized alpha-glucosidase can be achieved at protein concentrations up to 5 mg.ml^{-1}. In contrast, folding of the wild type enzyme in solution requires the much lower concentration of 15 *micro*grams per milliliter.[50]

References

1. Mosbach K, ed. Immobilized Enzymes. Meths Enzymol 44. NY: Academic Press Inc., 1976.
2. Mosbach K, ed. Immobilized Enzymes Part B. Meths Enzymol 135. NY: Academic Press Inc., 1987.
3. Mosbach K, ed. Immobilized Enzymes Part C. Meths Enzymol 136. NY: Academic Press Inc., 1987.
4. Woodward J, ed. Immobilized Cells and Enzymes: a Practical Approach. Oxford: IRL Press, 1985.
5. Bickerstaff GF. Applications of immobilized enzymes to fundamental studies on enzyme structure and function. In: Wiseman A, ed. Topics in Enzyme and Fermentation Biotechnology, vol 9. Chichester UK: Ellis Horwood, 1984: 162-201.
6. Martinek K, Mozhaev VV. Immobilization of enzymes: an approach to fundamental studies in biochemistry. In: Meister A, ed. Adv Enzymol 57. NY: John Wiley & Sons, 1985:179-249.
7. Katchalski-Katzir E. Immobilized enzymes—learning from past successes and failures. Trends Biotechnol 1993; 11: 471-478.
8. Chang TMS. Stabilization of enzymes by microencapsulation with a concentrated protein solution or by microencapsulation followed by cross-linking with glutaraldehyde. Biochem Biophys Res Comm 1971; 44:1531-1536.
9. Johnson DB. The stability of immobilized enzymes. Biochem Soc Transact 1979; 7:7-11.
10. Norde W. The behavior of proteins at interfaces in relation to their structural stability. In: van den Tweel WJJ, Harder A, Buitelaar RM, eds. Stability and Stabilization of Enzymes. Amsterdam: Elsevier, 1993:3-11.
11. Martinek K, Klibanov AM, Goldmacher VS et al. The principles of enzyme stabilization I. Increase in thermostability of enzymes covalently bound to a complementary surface of a polymer support in a multipoint fashion. Biochim Biophys Acta 1977; 485:1-12.
12. Martinek K, Klibanov AM, Goldmacher VS et al. The principles of enzyme stabilization II. Increase in the thermostability of enzymes as a result of multipoint

noncovalent interaction with a polymeric support. Biochim Biophys Acta 1977; 485:13-28.
13. Gabel D. The denaturation by urea and guanidinium chloride of trypsin and N-acetylated-trypsin derivatives bound to Sephadex and agarose. Eur J Biochem 1973; 33:348-356.
14. Mozhaev VV, Siksnis VA, Torchilin VP et al. Operational stability of copolymerized enzymes at elevated temperatures. Biotechnol Bioeng 1983; 25:1937-1945.
15. Mozhaev VV, Melik Nubarov NS, Sergeeva MV et al. Strategy for stabilizing enzymes, part one: increasing stability of enzymes via their multi-point interaction with a support. Biocatalysis 1990; 3:179-187.
16. Tadzhiev MKh, Ibragimov F, Khasanov KhT et al. Immobilization and stabilization of enzymes by polymers with functional epoxy groups. Appl Biochem Microbiol 1978; 14:544-548.
17. Blanco RM, Calvete JJ, Guisán JM. Immobilization-stabilization of enzymes; variables that control the intensity of the trypsin(amine)-agarose (aldehyde) multipoint attachment. Enzyme Microb Technol 1989; 11:353-359.
18. Blanco RM, Guisán JM. Stabilization of enzymes by multipoint covalent attachment to agarose-aldehyde gels. Borohydride reduction of trypsin agarose derivatives. Enzyme Microb Technol 1989; 11:360-366.
19. Torchilin VP, Maksimenko AV, Smirnov VN et al. Principles of enzyme stabilization V. The possibility of enzyme selfstabilization under the action of potentially reversible intramolecular cross-linkages of different length. Biochim Biophys Acta 1979; 568:1-10.
20. Trubetskoy VS, Torchilin VP. Artificial and natural thermostabilization of subunit enzymes: do they have similar mechanism? Int J Biochem 1985; 17:661-663.
21. Guisán JM, Fernández-Lafuente R, Rodriguez V et al. Enzyme stabilization by multipoint covalent attachment to activated pre-existing supports. In: van den Tweel WJJ, Harder A, Buitelaar RM, eds. Stability and Stabilization of Enzymes. Amsterdam: Elsevier, 1993:55-62.
22. Fernández-Lafuente R, Rosell CM, Alvaro G et al. Additional stabilization of penicillin G acylase-agarose derivatives by controlled chemical modification with formaldehyde. Enzyme Microb Technol 1992; 14:489-495.
23. Misset O. Stability of industrial enzymes. In: van den Tweel WJJ, Harder A, Buitelaar RM, eds. Stability and Stabilization of Enzymes. Amsterdam: Elsevier, 1993:111-131.
24. Sauber K. Lessons from industry. In: van den Tweel WJJ, Harder A, Buitelaar RM, eds. Stability and Stabilization of Enzymes. Amsterdam: Elsevier, 1993: 145-151.
25. Shewale JG, Deshpande BS, Sudhakaran VK et al. Penicillin acylases: applications and potentials Process Biochem Int 1990; 25:97-102.
26. Valle F, Balbás P, Merino E et al. The role of penicillin amidases in nature and in industry. Trends Biochem Sci 1991; 16:36-40.
27. Virden R. Structure, processing and catalytic action of penicillin acylase. Biotechnol Genet Eng Rev 1990; 8:189-218.
28. Lindsay C, Pain RH. The folding and solution conformation of penicillin G acylase. Eur J Biochem 1990; 192:133-141.
29. Alvaro G, Fernandez-Lafuente R, Blanco RM et al. Immobilization-stabilization of Penicillin G acylase from *Escherichia coli*. Appl Biochem Biotechnol 1990; 26:181-195.
30. Alvaro G, Fernandez-Lafuente R, Rosell CM et al. Penicillin G acylase from *Kluyvera citrophila*: new choice as industrial enzyme. Biotechnol Lett 1992; 14:285-290.

31. Guisán JM, Alvaro G, Fernandez-Lafuente R et al. Stabilization of heterodimeric enzyme by multipoint covalent immobilization: penicillin G acylase from *Kluvera citrophila*. Biotechnol Bioeng 1993; 42:455-464.
32. Bianchi D, Golini P, Bortolo R et al. Immobilization of penicillin G acylase on aminoalkylated polyacrylic supports. Enzyme Microb Technol 1996; 18:592-596.
33. Sheehan H, O'Kennedy R, Kilty CG. Investigation of the properties of bovine heart creatine kinase cross-linked with dimethylsuberimidate. Biochim Biophys Acta 1990; 1041:141-145.
34. Rudge J, Bickerstaff GF. Thermal stability propeties of immobilized creatine kinase. Biochem Soc Transact 1984; 12:311-312.
35. Gianfreda L, Greco G. The stabilizing effect of soluble macromolecules on enzyme performance. Biotechnol Lett 1981; 3:33-38.
36. Lenders JP, Germain P, Crichton RR. Immobilization of a soluble chemically thermostabilized enzyme. Biotechnol Bioeng 1985; 27:572-578.
37. Coulet P, Julliard JH, Gautheron DC. A mild method of general use for covalent coupling of enzymes to chemically activated collagen films. Biotechnol Bioeng 1974; 16:1055-1068.
38. Moreno JM, Ó Fágáin C. Immobilization of alanine aminotransferase by covalent binding and entrapment. Biotechnol Appl Biochem 1996; 23:231-235.
39. Moreno JM, Ó Fágáin C. Stabilization of alanine aminotransferase by consecutive modification and immobilization. Biotechnol Lett 1996; 18:51-56.
40. St. Clair NL, Navia MA. Crosslinked enzyme crystals as robust biocatalysts. J Am Chem Soc 1992; 114:7314-6.
41. Navia MA, St Clair NL. Use of crosslinked crystals as a novel form of enzyme immobilization. Patent Cooperation Treaty Application WO92/02617: 1992.
42. Navia MA, St Clair NL, Griffith JP. Crosslinked enzyme crystals (CLECs™) as immobilized enzyme particles. In: van den Tweel WJJ, Harder A, Buitelaar RM, eds. Stability and Stabilization of Enzymes. Amsterdam: Elsevier, 1993:63-73.
43. Simon LM, Kotorman M, Szajani B et al. Preparation and characterization of immobilized glucose-phosphate isomerase. Enzyme Microb Technol 1986; 8:222-226.
44. Mozhaev VV, Berezin IV, Martinek K. Reactivation of immobilized enzymes. Meths Enzymol 1987; 135:586-596.
45. Hagen AJ, Hatton TA, Wang DIC. Protein refolding in reversed micelles. Biotechnol Bioeng 1990; 35:955-965.
46. Hagen AJ, Hatton TA, Wang DIC. Protein refolding in reversed micelles: interactions of the protein with micelle components. Biotechnol Bioeng 1990; 35:966-975.
47. Cho IC, Swaisgood H. The reactivation of an unfolded subunit enzyme covalently linked to a solid surface. Biochim Biophys Acta 1972; 258:675-679.
48. Chan WWC, Mosbach K. Effects of subunit interactions on the activity of lactate dehydrogenase studied in immobilized enzyme systems. Biochemistry 1976; 15:4215-4222.
49. Stempfer G, Höll-Neugebauer B, Kopetzki E et al. A fusion protein designed for noncovalent immobilization: stability, enzymatic activity, and use in an enzyme reactor. Nature Biotechnology 1996; 14:481-484.
50. Stempfer G, Höll-Neugebauer B, Rudolph R. Improved refolding of an immobilized fusion protein. Nature Biotechnol 1996; 14:329-334.

Chapter 8

Protein Engineering

Introduction

The term 'protein engineering' refers to the use of genetic manipulation techniques to alter proteins of interest in extremely specific ways. Substitution of a single amino acid residue at a known location can be accomplished routinely by site-specific (or site-directed) mutagenesis protocols. This approach has already had a major impact on our knowledge of protein structure and function. A detailed consideration of the strategies and techniques involved in protein engineering is beyond the scope of this chapter. A number of informative reviews have appeared, however.[1-4] References 5-7 may be cited among available monographs. Useful descriptions of some of the key techniques of protein engineering appear in ref. 6.

Site-directed mutagenesis depends on the use of synthetic oligonucleotides to direct the required mutation. It is highly specific: individual bases may be reliably altered even within a triplet codon. Examples of the sorts of substitutions possible include replacement of tyrosine by phenylalanine to assess the contribution of the tyrosine -OH group, or exchange of aspartic with glutamic acid to extend or shorten the position of the $-COO^-$ function by a distance of one $-CH_2-$ group. One can also insert non-standard amino acids (such as L-3-iodotyrosine,[8] (S)-p-nitrophenylalanine, homophenylalanine or (R)-3-amino-2-benzylpropionic acid[9]) at desired locations.[10] Non-natural amino acid substitutions have already been engineered specifically to investigate protein stability.[11]

To undertake site-directed mutagenesis, a synthetic oligonucleotide (incorporating the desired mutation) is annealed to a single-stranded gene copy (produced by, for example, use of the bacteriophage M13). Extension of the oligonucleotide occurs from the 3' end. Wild type and mutant phenotypes will be expressed on subsequent infection of a bacterium. The aptly-named cassette mutagenesis involves the insertion of a double-stranded segment of DNA into a suitable DNA vector between two non-identical restriction sites. The resulting recombinant DNA contains no mismatches and results in highly efficient mutagenesis since DNA repair does not occur.[12] The polymerase chain reaction (PCR) has become widely used to create mutant protein forms.[13,14] The two-codon insertion strategy should also be mentioned. This is based on the hypothesis that, in an 'unknown', novel or poorly-characterized protein, the activity of insertion mutants will vary according to the insertion's three-dimensional distance from the active site. One can readily see that substitution of those active site amino acids involved in the catalytic mechanism will adversely affect enzyme activity. Substitution of non-catalytic residues

Stabilizing Protein Function, by Ciarán Ó Fágáin.
© Springer - Verlag and Landes Bioscience 1997.

close to the active site, which function to position the catalytic residues correctly, is less likely to be deleterious. However, *insertion* mutations among these 'positioning' residues will likely be destructive, due to changes in the relative orientations of the catalytic amino acids. Positions more remote from the active site will be more tolerant of insertions while positions within loops' linking domains may be unaffected by insertions. Accordingly, the two-codon insertion methodology may help identify active site or functional domains in a protein of unknown structure. The severity of effect of a two-codon insertion lies between that of a point mutation and a large deletion, so that insertion is a feasible means of investigating structure-function relationships.[15]

This ability to create a clearly-defined mutant protein has allowed the purposeful manipulation of stability by protein engineering (see refs. 5-7). Many proteins have already been altered in this way, including lysozyme (see below and refs. 16-19), cytochrome c[20] and firefly luciferase.[21] Here the focus will be on just a few prominent examples, including the proteolytic enzyme subtilisin and the sugar-transforming glucose isomerase, both of which have industrial applications.

Ground Rules for Protein Engineering

Imanaka and co-workers have suggested a route towards engineering proteins with enhanced thermostability.[22] Highly conserved residues should not be altered. One should use known sequences for corresponding proteins from mesophiles and thermophiles to work out the sorts of changes required for increased thermostability. Suitable substitutions are those which will increase the protein's internal hydrophobicity and will increase the strength of helical internal packing. Imanaka's group applied this strategy to a thermostable *Bacillus stearothermophilus* neutral protease, describing in detail the mutation techniques employed. A mutation G144A (where the wild type glycine residue 144 is replaced by alanine; letters indicate the single-letter amino acid code, numbers the position in the polypeptide sequence) gave increased stability versus the wild type protein but a T66S mutant was less stable. Double mutants (G144A-T66S and G61A-T66S) were as labile as the T66S single mutant. Only construction of a triple mutant (G144A-G61A-T66S) permitted some recovery of stability in the engineered protease. However, this triple mutant's stability did not match that of the original wild type protein.[22]

Menendez-Arias and Argos have suggested an alternative means of engineering protein thermostability.[23] In a statistical survey of the thermophilic and mesophilic sequences of six different protein homologues (lactate dehydrogenase, glyceraldehyde-3-phosphate dehydrogenase, triose phosphate isomerase, superoxide dismutase, thermolysin and subtilisin), they found that alpha-helical regions consistently showed lesser flexibility and greater hydrophobicity in adapting to higher temperatures. In thermophiles, domain interfaces also seem to display this decreased flexibility and increased hydrophobicity. Based on these findings, the authors proposed empirical rules to improve the stability of natural or engineered proteins. Where information on the tertiary structure is available, one should focus on the helices. Where only the primary structure is known, one should use secondary structure prediction methods to ascertain likely helical regions. One can then devise suitable amino acid substitutions to decrease flexibility and increase hydrophobicity (e.g. Gly or Ser $\rightarrow$ Ala, Lys $\rightarrow$ Arg or Ala: a full range of possible substitutions is given in ref. 23). Steric interferences and large/small substitutions should be avoided to preserve structural integrity and notable features such as ligand binding sites. Note that one can use this stabilization strategy in the absence of a three-dimensional structure. Interestingly, Menendez-Arias and Argos point out that most of the stabilizing amino acid changes lead to increased (G+C) content in the coding DNA. (G+C) content correlates with DNA thermal stability, echoing Zuber's idea (ref. 24 and chapter 4) that stabilizing amino acid substitutions correlate with a more thermostable DNA.

Noting that existing theoretical methods for free energy stability calculations were cumbersome, Lee and Levitt perceived a need for theoretical stability prediction methods sufficiently simple and fast for testing of a large set of mutations and not requiring parameter adjustment to fit each individual mutant. These scientists have reported a theoretical method for the calculation of the stabilizing energies of mutant proteins, based on experimental data for a set of lambda repressor mutants. The calculated energy values correlated closely (92% accuracy) with the activities of the various mutants and even more closely with their thermostabilities. This method was based on their own earlier theoretical framework for the prediction of side-chain conformations in protein cores and included some simplifying assumptions, which were borne out by the excellent correlation between predicted and experimentally-determined values. The calculated energy for each mutant is denoted E_{calc}; Lee and Levitt postulated that mutants with higher E_{calc} values than wild type lambda repressor should be less active due to active site distortion and destabilization of the folded state. Active and inactive mutants clustered in two distinct energy ranges with very little overlap. Interestingly, E_{calc} values correlated with activities even though E_{calc} is primarily a stability predictor. However, some cases did not illustrate a clear relation between activity and stability: one triple mutant (V36L/M40L/V47I), was more stable than the wild type (and had the predicted lower E_{calc}) but was less active. A noteworthy feature of the method is its ability to predict stability changes in the mutant cohort in spite of the model's complete neglect of the unfolded state.[25]

Nosoh and Sekiguchi considered protein stabilization factors in more general terms.[4] One can change a protein's stability by substituting:

i) amino acids involved in stabilizing interactions in the folded protein,
ii) those involved in destabilizing interactions in the unfolded form, or
iii) both.

One can group these interactions into three categories:

- bond factors (e.g. hydrogen bonds and hydrophobic interactions)
- conformational factors (such as secondary structure stability or internal hydrophobicity) and
- protection factors (e.g. protection of sulfhydryl group oxidation).

It is appropriate at this point to survey some examples of enzyme stabilizations achieved by protein engineering. Of particular interest are the results obtained for the enzymes lysozyme, subtilisin, triose phosphate isomerase, glucose isomerase and kanamycin nucleotidyl transferase.

Case Study: Lysozyme

Lysozyme functions in the breakdown of bacterial cell walls. It is a small, well characterized enzyme and the X-ray crystal structure of the enzyme from bacteriophage T4 has been determined.[26] (See Fig. 8.1.) Ile is present at position 3 of the wild type T4 lysozyme. Site-directed mutagenesis of this residue, followed by oxidation in vitro, led to formation of a new crosslink between the introduced Cys 3 and the naturally-occurring Cys 97. This new disulfide bond stabilized the enzyme against both irreversible thermal inactivation[27] and against reversible thermal unfolding in the pH range 1-7.[28] Resistance to irreversible thermal inactivation at 67°C increased further following replacement in the disulfide-crosslinked mutant of the

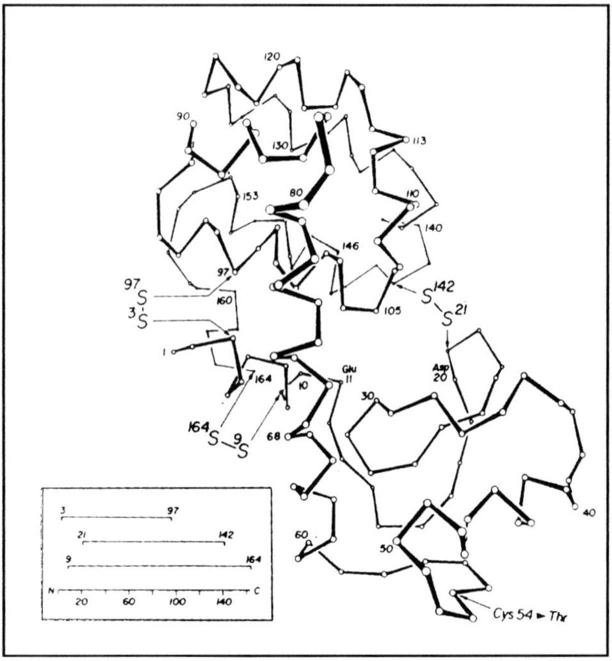

Fig. 8.1. Crystal structure of T4 lysozyme. Reprinted with permission from: Matsumura M, Signor G, Matthews BW. Nature 1989; 342:291. © 1989 Macmillan Magazines Ltd.

unpaired wild type Cys 54 by threonine. This suggests that the unpaired cysteine residues are prone to oxidative degradation (chapter 2).

Two double-disulfide mutants (3-97/9-164 and 9-164/21-142) were likewise prepared, together with a triple mutant incorporating all three bridges. (Note that the 21-142 disulfide abolishes catalytic activity: the crosslink spans the 'mouth' of the active site cleft.) Both of the double mutants were more stable than the corresponding single mutants. The T_m melting temperature for the triple mutant was 23.4°C higher than that of the wild type. This value agreed well with the summed T_m increases for the corresponding single mutants.[17,29] The authors recommend that disulfide bridges introduced to stabilize a protein should be designed to form large loops. This will maximize the entropic effect on the unfolded protein. One should also choose flexible sites to avoid the possibility of strain in the folded protein.

Matthews and colleagues altered Gly 77 and Ala 82 of wild type lysozyme to Ala and Pro,. respectively. Stability of this derivative increased by 3.6 kJ.mol^{-1} (ref. 30). Replacement of the glycine by any other residue was stabilizing. Substitution of proline for different residues was also stabilizing. This agrees with Suzuki's observation that thermostable proteins have a greater proline content, leading to a more rigid tertiary structure.[31] Nicholson and colleagues noted that there is an effective positive charge at the N-terminus of an alpha-helix and a negative charge at its C-terminus. Introduction of a negatively-charged aspartic acid residue at the N-terminus of an alpha-helix (Ser 38 or Asn 144 → Asp) yielded a stabilized derivative. The double mutant had twice the increase in stability associated with the single mutants.[32]

Matsumura and colleagues focused on the Ile3 residue of T4 lysozyme and replaced it with other residues. Among the various substitutions, changes in thermodynamic stability coincided with the hydrophobicity of the replacement residue. Among these residues, only Val increased the protein's T_m value (by 3.0°C and 0.9°C at pH 2.0 and 6.5, respectively). (Cysteine at position 3 was destabilizing in the reduced state but could be induced to form a disulfide with Cys97; if present, this new disulfide increased T_m by 4.8° and 3.3°C at pH 2.0 and 6.5, respectively.) Examination of the molecular structure indicated that hydrophobic stabilization led to a decreased solvent-accessible surface area on folding.[33]

It has also proven possible to combine a number of individual stabilizing mutations within the same T4 lysozyme variant to achieve a cumulative enhancement of stability. Seven rationally designed point mutations were engineered into the same lysozyme molecule, leading to an 8.3°C increase in T_m and a 3.6 kcal.mol^{-1} change in Gibbs energy. The substitutions were S38P, A82P, N144D, I3L, V131A, A41V and N116D. (The variant incorporating all of these changes was dubbed 'SANIVAN' from the single letter designations of the wild type residues.) Stability measurements took place at pH 3.0 and 5.4. The stability gains for multiple mutants equalled the sum of the individual stability increments, indicating that each mutation was independent of the others and that any structural alterations were confined to the immediate vicinity of the substitution. The unfolding Gibbs energies of mutations which conserved the enzyme molecule's electrical charges were independent of pH, as one would expect. Charge-altering substitutions (S38D, N116D and multiple mutants involving these two replacements) displayed pH-dependent Gibbs energies of unfolding. These two stabilizing alterations (and N144D) are each thought to interact with an alpha helix dipole. One would expect this sort of interaction to strengthen with increasing pH from 3.0 to 5.4 (see discussion of lysozyme and barnase in chapter 2). The purpose of the A82P and A41V changes was to decrease the entropy of the unfolded state while the V131A mutant is thought to favor alpha helix formation. The I3L substitution probably lessens exposure of hydrophobic surface area in the folded enzyme and may also relieve some strain present in the wild type. Thus mutations yielding stability gains by different mechanisms can be combined to have a cumulative beneficial effect on lysozyme stability. The protein structure underwent only small local changes in gaining this extra stability (notably in the regions 67-71 and 97-100). However, there is a significant hinge bending displacement, which is probably attributable to the I3L mutation alone. However, one of the stabilizing substitutions, N116D, leads to a notable decrease in activity: the N116 residue is involved in substrate binding. Also, lytic activity reduces with the accumulation of the other six mutations.[34] This echoes the case of the 21-142 disulfide mutant above where the price of stabilization was the loss of catalytic activity and highlights once again the compromise between catalytic efficiency and molecular stability that is a feature of many enzymes (outlined in chapters 2 and 4).

The T4 enzyme is not the only lysozyme that has been stabilized by protein engineering. Malcolm and colleagues describe studies on wild type and site-specific mutant forms of lysozyme among game birds (Order *Galliformes*).[16] Activities, thermostabilities and three-dimensional structures were each examined. Attention focused on the residues at positions 40, 55 and 91. In all known game birds, Thr or Ser occupy position 40 while Ile or Val occupy position 55. Ser or Thr fill position 91. All three residues are closely located in the three-dimensional structure beneath the active site within the hinge between the two globular domains. The replacements at each position are notably conservative: Ser and Thr each have an -OH function in their side chains while Ile and Val are both aliphatic, hydrophobic amino acids. Malcolm et al.postulated the existence of intermediates between the Thr/Ile/Ser state and the Ser/Val/Thr variant. They assessed thermostability by measurement of the melting temperature T_m and also by use of an Arrhenius plot (logarithm of experimental inactivation rate constant versus reciprocal of absolute temperature; see chapter 1) to calculate residual activity after incubation for one week at 37°C. T_m for the two wild type configurations lay between 73° and 74°C. T_m values correlated strongly with the identities of the residues in positions 40, 55 and 91. Replacement of Ser by Thr at position 91 to give a Thr40/Ile55/Thr91 triplet yields a stabilized lysozyme. This results from the introduction of a new methyl group which partially fills a cavity in the wild type enzyme. The lysozyme molecule's ability to absorb this extra group without strain may explain this variant's enhanced thermostability. Note that this change also leads to increases in packing density and internal hydrophobic contacts. Thus, it appears that one can increase a protein's thermostability by simple hydrophobic packing in accordance with the observations of Kellis et al and Yutani et al.[35,36] Interestingly, the Arrhenius procedure suggested that differences in residual catalytic activity may arise between the various mutants during the period of egg incubation.[16]

Tomizawa et al established that the irreversible thermoinactivation of hen egg lysozyme was due to an accumulation of chemical reactions, which progressively decreased the Gibbs energy difference between the folded and unfolded states.[37] (Indeed, some lysozyme molecules unfolded and lost activity even at low temperatures.) Prominent among these reactions were the isomerization of an Asp-Gly sequence and deamidation of an Asn residue. Asp and Asn residues with Gly on their C-terminal side are especially subject to their respective destructive reactions.[37-39] (See also chapter 2). Tomizawa et al reasoned that substitution of these labile residues should suppress the deleterious reactions. Glycines in the three hen egg lysozyme Asp-Gly pairs at positions 48-49, 66-67 and 101-102 were replaced with alanines. The 101-102 Asp-Gly pair is immediately followed by an Asn-Gly pair at positions 103-104 and Gly 104 was likewise replaced by Ala. An insertion sequence of six human lysozyme residues (Tyr-Asn-Ala-Gly-Asp-Arg) was also substituted for the five hen egg positions 45-49 (Arg-Asn-Thr-Asp-Gly). Activities of individual Gly→Ala mutants closely resembled that of the wild type but the insertion mutant was somewhat less active. (Note that variants containing an unsubstituted Gly102 undergo an initial activation of lytic activity at high temperatures, due to formation of a highly reactive cyclic imide at Asp101-Gly102.) All mutants (except the G102A-G104A double mutant) were more resistant to irreversible inactivation of lytic activity at 100°C and pH 4; the inactivation rate constant of the most stable mutant was half that of native lysozyme. Curiously, however, the stabilization of the mutants against irreversible inactivation was accompanied by a *lower* reversible conformational stability (i.e. more prone to denaturation by guanidine hydrochloride). This was explained by the fact that Gly49, Gly67 and Gly102 are involved in a turn and a loop and that the larger Ala R-groups in these positions destabilize the folded form. Multiple substitutions led to additive Gibbs energy changes, implying that the sites concerned were independent of each other.[38] These findings imply that the mutants unfold more easily than native lysozyme ($N \leftrightarrow U$ transition in the model in chapter 1) but that unfolded mutant lysozyme molecules are more resistant to irreversible inactivating events (the $U \rightarrow I$ portion of the model in chapter 1). Thus, a measurable improvement in lysozyme's resistance to inactivation has resulted from a stabilization of the unfolded enzyme rather than from the prevention of unfolding. These results also underline the importance of choosing appropriate stability indices (as discussed in chapter 1), since the favorable mutants could easily have been dismissed if attention had been confined to conformational criteria of stability only.

Case Study: Subtilisin

Subtilisin is an exoenzyme produced by *Bacillus* species. It is used in detergent formulations and for selective biocatalysis.[40] It is a single-domain, endo-acting serine proteinase of low specificity and with a molecular weight of 27,500. No disulfide bonds occur in the native structure. It is very well characterized: its gene has been cloned, sequenced and expressed at high level and many X-ray crystal structures have been solved. All these features make subtilisin an ideal candidate for protein engineering experiments.[40,41] Common variants include subtilisin BPN' from *B. amyloliquefaciens*, subtilisin Carlsberg from *B. licheniformis* and Savinase and Esperase (both registered trademarks) from the highly alkalophilic *B. lentus*.[42]

Despite their lability under certain conditions.[43-45] (See also chapter 2), disulfide bonds act as a natural intramolecular crosslink which will stabilize the folded form of a protein. Since it lacks cysteines, at least two such residues (separated by an appropriate distance in the three-dimensional structure) must be substituted into subtilisin in engineering a disulfide crosslink. Wells and associates formed a disulfide bond in vitro between the introduced Cys22 and Cys87. They also prepared a disulfide derivative with cysteines at positions 24 and 87. These introduced disulfides did not notably stabilize subtilisin against irreversible inactivation (N → I; chapter 1); indeed, the Cys22-Cys87 mutant was less stable than the wild type. Further disulfide variants were produced but proved no more stable than the wild type against irreversible inactivation. The possible disruption of existing stabilizing interactions within the protein by the introduced disulfides may explain these findings.[46,47] Five different -S-S- bridges were introduced into subtilisin by substituting cysteine residues to give single disulfide bridges at positions 26-232, 29-119, 36-210, 41-80 and 148-243. None of these mutants was substantially more stable than the wild type BPN' enzyme. Some of the mutants were much more stable than their reduced counterparts (no disulfides), however.[46]

A somewhat different strategy did lead to stabilizing disulfide mutants of subtilisin. A different research group did manage to achieve stable disulfide mutants, but not until three generations of mutants had been prepared. Disulfide insertion sites were selected in three distinct steps: first, a high-resolution three-dimensional subtilisin structure was inspected by computer. Locations were noted where the atomic positions of two main chain residues closely resembled those of main chain atoms of known disulfides from a protein data bank. Second, computer graphic techniques were used to insert cysteines/sulfurs in the locations identified in step one. This allowed screening of the potential disulfide sites: unsuitable locations were eliminated. Third, sites known to be conserved in other alkaline proteases were not altered.[48] Only when this additional conservation criterion was adopted did stabilized mutants result. Three pairs of potential substitutions were identified by this strategy; two proved more conformationally stable (N ↔ U; chapter 1) than the wild type. These were T22C/S87C and Q206C/A216C; each had T_m values 3°C higher than the wild type's under nonreducing conditions. Both, however, were less stable than wild type in presence of 10 mM dithiothreitol, a reducing agent.[40,48] The stability gain observed was consistent with the entropy decrease for the unfolded protein derived from polymer statistical thermodynamics.[40] Also, Tagaki and colleagues successfully stabilized subtilisin E by engineering in Cys residues at positions 61 and 98 to form a disulfide bridge.[49]

Asparagine is a labile amino acid that is prone to deamidation and also to cyclization reactions at Asn-Gly pairs. (See chapter 2 and lysozyme case history in paragraphs above.) The aprA-subtilisin from *B. subtilis* contains Asn-Gly sequences at Asn109 and Asn218. Substitution of these two Asn residues by Ser, to yield the mutants N109S and N218S, led to an additive stabilization of the enzyme by 7°C, as measured by differential scanning calorimetry.[50]

Calcium ion stabilizes subtilisin and X-ray structures show that the enzyme has both a high affinity and a low affinity calcium binding site. Substitution of aspartic acids for Gly131 and Pro172 introduced additional negative charges near the low affinity binding site. This led to significant stability gains. Stability also increased when extra negative charges were introduced in the vicinity of the high affinity binding site; this was accomplished by a N76D substitution (replacement of Asn76 by Asp).[51] Exactly this change, the deamidation of Asn to Asp, has been implicated in the destabilization or inactivation of proteins[43] such as triose phosphate isomerase.[52] In the present case, the engineered additional negative charges lead to improved electrostatic interactions between the calcium ions and the subtilisin protein. The benefits of these enhanced interactions obviously outweigh any

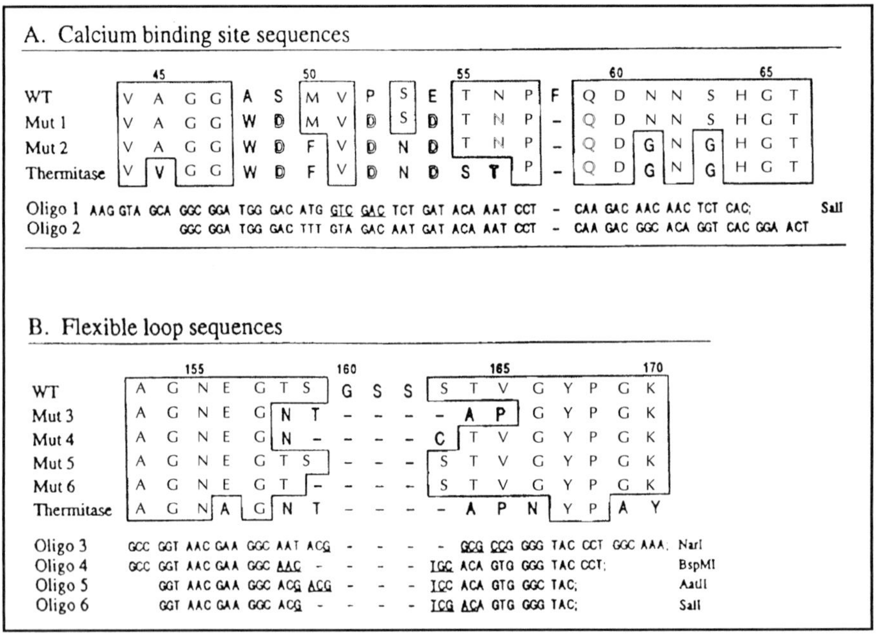

Fig. 8.2. Protein sequence comparison between subtilisin and thermitase spanning a Ca^{2+}-binding site (a) and a flexible loop region in subtilisin (residues 158-164) (b). Mutants 1 and 2 were designed to confer Ca^{2+} binding to subtilisin in a region susceptible to autolysis. Residues involved in coordination of calcium are Asp49, Asp52, Asp54, Thr56, and Gln59. Mutants 3-6 were were designed to alter a flexible loop surrounding position 164 that is susceptible to autolysis. The boxed sequences indicate sequence identity, and the open letters indicate sites of mutation. The oligonucleotides used in the mutagenesis are shown, and the underlined segments indicate new restriction sites. Reprinted with permission from: Braxton S, Wells JA. Biochemistry 1992; 31:7797. © 1992 American Chemical Society.

potential drawbacks of aspartic acid insertion. The identical N76D substitution also had a positive effect on the stability of subtilisin-aprA.[50] Bryan notes that combination of individual stabilizing mutations led to approximately additive increases in the Gibbs energy of stabilization but to multiplicative increases in the half-times for thermal inactivation.[53] Braxton and Wells successfully incorporated a third calcium binding site into subtilisin BPN'.[54] This was derived from a calcium-binding loop (around Ser49) occurring in thermitase, a thermophilic protease homologous with subtilisin. This calcium-binding sequence within thermitase corresponds to residues 45-63 in subtilisin BPN'. This 45-63 region of wild type subtilisin is a highly mobile extended surface loop which contains a prime site of autolytic digestion at Ala48-Ser49. The purpose of this molecular graft was to remove this site of autolysis so as to stabilize the resulting chimeric BPN' enzyme. Within the corresponding thermitase sequence, Asp 49, Asp 52, Asp54, Thr56 and Gln59 coordinate a calcium ion.[55] (See Fig. 8.2.) The grafted enzyme molecule could bind a third calcium ion with reasonable affinity (K_d was approximately 100 micromolar) and had the same k_{cat} value as wild type BPN', although its K_m value was 8 times larger. The chimeric enzyme was slightly less resistant to thermal inactivation in EDTA, using the chromogenic artificial substrate sAAPF-pNA (N-succinyl-Ala-Ala-Pro-Phe-p-nitroanilide) to measure activity. Possibly, this is due to the extra negative charge repulsion within the mutant molecule in the absence of calcium. In the presence of 10mM $CaCl_2$, however, the mutant was 10 times more stable than the wild type to irreversible inactivation at 60°C. The activity loss at 60°C was first order and independent of subtilisin concentration. Possible explanations for these kinetics include i) irreversible unfolding of a loop structure close to the active site, ii) an intramolecular event involving self-proteolysis of a subtilisin molecule and/or iii) a rate-limiting reversible local unfolding where the unfolded region is rapidly proteolyzed. Note that the stabilizing effects of calcium on the subtilisin molecule are pH-dependent. At pH 3.5, where Asp residues are not fully ionized and are therefore unable to complex calcium, neither the wild type nor chimeric BPN' enzymes are stabilized by calcium. Also, the observed stabilization seems to be due to increased resistance to autolysis and not to any increase in folding stability: T_m values in 10mM $CaCl_2$ for the wild type and chimeric BPN' enzymes were 73.4 and 72.0°C respectively. Braxton and Wells attempted to build on this success by grafting thermitase-derived mutations and deletions near to a second BPN' autolysis loop at residues 158-165. Unfortunately, all of these latter mutants were less stable than wild type subtilisin BPN'. It appears that factors other than primary structure alone influence autolysis in this region.[54]

Whatever its merits in nature, subtilisin's requirement for calcium for maximum stability is a disadvantage in industrial applications (such as detergents) where chelating agents occur in high concentration. What would happen if one were to abolish a calcium binding site altogether? Strausberg and colleagues set out to answer this question. The conserved calcium-binding loop Site A (residues 75-83) separates two alpha-helical regions spanning residues 63-85 (Fig. 8.3). Strausberg et al[56] deleted this calcium-binding loop to abolish calcium binding at site A and create a subtilisin variant with a continuous stretch of alpha-helix. This molecule, dubbed S46, was considerably less stable than the natural subtilisin (although its X-ray structure quite closely resembled that of wild type subtilisin). However, it could be (and was) used as a 'baseline' molecule for successive rounds of mutagenesis aimed at producing a stabilized form of the enzyme free from any calcium dependence. This was in fact accomplished in a three-step program. First, the

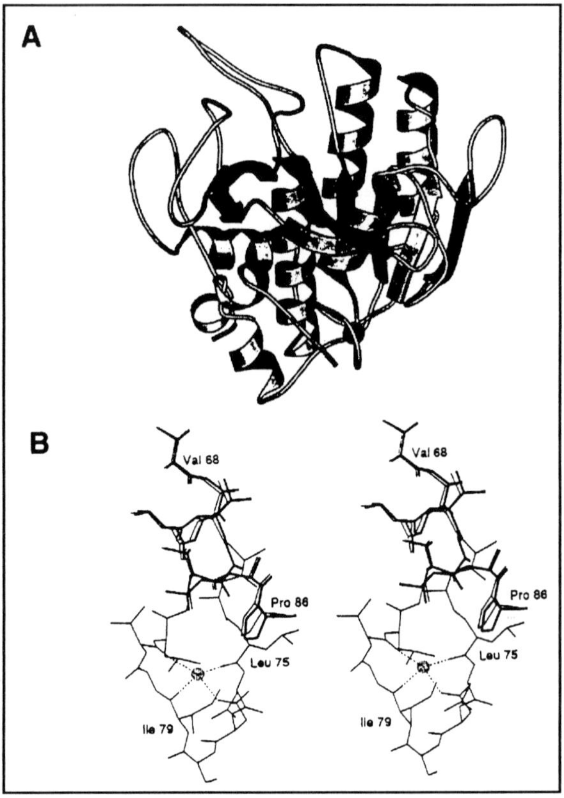

Fig. 8.3. Three-dimensional structure of subtilisin. (A) Ribbon diagram depicts the topology of the subtilisin BPN' backbone. Solid sphere shows the position of the high affinity calcium-binding site A. (B) Stereo superimposition of the continuous helix of Δ75-83 (S15) subtilisin (thick lines) onto the interrupted wild type helix (thin lines). In the wild type, four of the six ligands to calcium are provided by the carbonyl oxygens in the loop composed of amino acids 74-82. The helix of Δ75-83 subtilisin is uninterrupted and shows normal helical geometry over its entire length. The structures of subtilisin with and without the deletion superimpose with an rms difference between 261 C-alpha positions of 0.17Å. Residues 75-83 were chosen for deletion, as opposed to 74-82 (those actually belonging to the loop) out of preference for A74 rather than G83 in the resulting continuous helix. Alanine has a higher statistical likelihood for occurrence in an alpha-helix, due to glycine's broader range of accessible backbone conformations. Reprinted with permission from: Strausberg Sl, Alexander PA, Gallagher DT et al. Bio/Technology 1995; 13:670. © 1995 Nature Biotechnology.

structure and stability of the 'loop-less' S221C mutant was analyzed. (The S221C mutation was required to prevent autolysis of loop-deleted subtilisin variants. Also, the previously-identified stabilizing mutations M50F, Y217K and N218S[53] were introduced for the same reason. This inactive mutant subtilisin was termed S15.) Second, proteolytic activity was restored by the reverse mutation C221S. (Again, this procedure turned out to be more complicated than anticipated. The gene for the active, calcium-free subtilisin S46 proved lethal to *E. coli* and had to be fragmented into non-functional pieces which were manipulated and mutagenized in *E. coli*. The gene fragments were then reassembled in vitro and used to transform *B. subtilis*. In this way the baseline S46 mutant, which was slightly more active than the wild type BPN' enzyme on sAAPFpNA, was obtained.) Third, this variant with restored protease activity was subjected to successive rounds of directed mutagenesis with selection for increased stability. Mutagenesis was undertaken on the basis that those amino acids interacting with calcium in the wild type enzyme were no longer optimal in the 'loop-less' baseline mutant. Inspection of the inactive S15 mutant structure indicated that the deletion had most affected the following regions: the N-terminal residues 1-4 (all of these had formerly interacted with the deleted calcium loop), the loop at residues 36-44, the alpha-helix 63-85 and the beta-strand 202-219. Ten residues in these regions were subjected to mutagenesis. The outcome of this program of experiments was a mutant subtilisin with proteolytic activity close to that of the native but with 1,000-times greater stability under chelating conditions. This mutant, according to Strausberg and colleagues, shows that the naturally-evolved, calcium-mediated stabilization is not the sole answer to the stability requirements of subtilisins.[56] Native subtilisin BPN' and the S46 calcium-free baseline mutant were thermoinactivated at 55°C in the presence of 10 mM EDTA. Half lives were 2 min and 25 min, respectively. Mutations in the N-terminal region had moderate effects. The Q2K and Q2L mutations were stabilized by 2-fold and 1.8-fold, respectively at 65°C with respect to the S46 form. The most dramatic stabilization (17-fold at 65°C) resulted from the spontaneous formation of a disulfide bridge in the S3C-Q206C double mutant. Reduction of this -S-S- link with dithiothreitol reversed the stabilization effect. This 'new' 3-206 disulfide is peculiar to 'loop-less' mutants of subtilisin. In the wild type protein, the presence of the 75-83 calcium-binding loop would prevent formation of a disulfide between the N-terminal portion and the 202-219 beta-ribbon. Substitutions in the 36-44 omega loop had modest effects, e.g., D41A was 1.5-fold more stable than S46. Changes in the 63-85 alpha helix and 202-220 beta-ribbon were more effective: the A73L variant had a 2.6 times longer half life than S46 while the Q206V and Q206I mutants were 4.5 and 4.0 times, respectively more stable than S46. (These stabilization factors are, of course, less than that obtained with the disulfide-forming S3C-Q206C double mutant.) The screening method involved incubation of individual *Bacillus* transformants at 37°C for 18 hours followed by a transfer step, incubation at 65°C for one hour and cooling to room temperature. Aliquots of the cooled samples were then assayed for activity against the sAAPF-pNA substrate. This procedure permitted identification of stabilizing mutations at six of the ten positions targeted. Slightly-stabilizing mutations were recorded at positions 5, 41 and 43 while notably-stabilizing mutations were identified at positions 2, 73 and 206. Further mutants containing combinations of these individual stabilizing mutations were constructed. One such variant, S88, combined the six mutations Q2K, S3C, P5S, K43N, A73L and Q206C. This S88 form of subtilisin displayed remarkable stability characteristics in comparison with the wild type BPN'. At 65°C in the presence of EDTA, S88 had a half life of 38 minutes while BPN' inactivated instantaneously. Even under non-chelating conditions and in the presence of excess calcium, S88 had a 1.5 times longer half life than BPN' at 65°C (90 minutes versus 60 minutes, respectively). The longer half life of S88 in the presence of calcium compared with chelating conditions is due to calcium binding to the second, weaker calcium binding site in the subtilisin molecule (see above). The two enzymes closely resembled each other in their kinetic parameters: k_{cat}/K_m values for the substrate sAAPF-pNA were 125,714 $M^{-1}.sec^{-1}$ and 155,319 $M^{-1}.sec^{-1}$ for BPN' and S88 respectively. These studies by Strausberg et al[56] are highly significant in that they have achieved an enzyme variant with activity similar to that of its wild type but with enhanced, calcium-independent stability and which can be obtained in high yield.

A group at Novo Nordisk (Denmark) prepared a subtilisin-variant (Savinase™)[57] insertion mutant, *36D, by inserting an Asp residue at position 36 of the alkalophilic *Bacillus lentus* strain's subtilisin. The X-ray structure of the *B. lentus* enzyme has been determined[58] and site-directed mutagenesis experiments have been described[59] using both double-stranded plasmids[60] and the polymerase chain reaction (PCR).[61] The rationale for this mutation was to lower the subtilisin's isoelectric point towards the pH of a liquid detergent formulation. However, the mutation also enhanced Savinase's thermostability by 4.0°C and led to significantly greater stability in liquid detergent. (Asp naturally occurs at position 36 in the homologous subtilisin BPN'.) However, the identical *36D insertion had no effect on the detergent-stability of the closely-related Esperase™ subtilisin variant and marginally destabilized it against temperature. Considering the sequences between positions 28-42 (BPN' numbering), Savinase and Esperase differ only at positions 37 and 38: Savinase™ has Ser and Thr, while Esperase™ has Ala and Ser, in these two positions. The stabilizing effect is 'non-transferable' from Savinase™ to Esperase™, despite their overall similarity in terms of primary structure. It is likely that this non-transferability is due to some structural feature or interaction at the tertiary level of protein structure.[42]

Suzuki, as noted above, has suggested the use of judicious proline substitutions to increase enzyme thermostabilities.[31] The Novo Nordisk scientists applied this principle to Savinase™, which already contains 13 proline residues.[62] In targeting particular sites for Pro substitution, the researchers took careful account of the dihedral angles of the candidate amino acid residues, since the rigid pyrollidine ring comprising the proline R-group greatly restricts the phi and psi dihedral angles which can be taken up. Also, no positions forming part of any alpha-helical or beta-sheet secondary structural elements were substituted, in order to avoid any possibly-destabilizing interferences with secondary structure hydrogen bonding patterns. On this basis, twelve mutants were constructed.[62] Three of these mutants (A160P, W241P, S265P) expressed too poorly in the recombinant *B. subtilis* strain to permit purification. This was ascribed to space problems and, possibly, to disruption of hydrogen bonds to main chain N-atoms resulting from the Pro substitutions.[62] The two mutants S24P and Q182P were less stable than the wild type: the alterations in their differential scanning calorimetry (DSC) transition temperatures were -2.0° and -2.5°C, respectively. The further variants A172P, A98P and T38P did not differ significantly from the wild type's DSC transition value. The remaining four mutants were stabilized with respect to the wild type by 0.6° (S259P), 0.7° (S57P), 1.5° (S188P) and 2.6°C (A194P). The A194P and S188P substitutions occurred at position 2 of a beta-turn and within a narrow range of phi/psi limits. These two mutants maintained more activity than wild type Savinase™ during storage stability tests in a U.S.-style (no bleach) liquid detergent: in general, DSC denaturation temperatures and storage stabilities correlated well. However, while the A98P change was neutral with respect to DSC denaturation temperature, this variant gave very poor storage stability results. The S57P change also resulted in poor storage stability characteristics, despite this substitution's 0.7°C increase in DSC temperature. It is possible that these two substitutions generated new sites for autoproteolysis. Such a development would have a much greater effect on long-term storage tests than on melting temperatures measured over a short period. Effects on the kinetics of sAAPF-pNA hydrolysis were minimal; only small variations in k_{cat} and K_m occurred among the eight mutants studied. The greatest difference between the k_{cat}/K_m values of the wild type Savinase™ and any mutant amounted to only a two-fold decrease. Interestingly, this was for the two variants A98P and S57P, which gave contradictory DSC and storage stability results. The slightly-stabilizing S57P and S259P changes occurred in beta-turns but not at position two of the turn, showing that 'broad range' phi-psi angles ($-90° < phi < -40°, 120° < psi < 180°$ OR $-90° < phi < -40°, -50° < psi < 10°$) can pinpoint sites for stabilizing Pro substitutions.[62]

The greatest industrial use of subtilisin is in detergents but the enzyme is prone to oxidative inactivation. A methionine residue M222 lies next to the essential serine

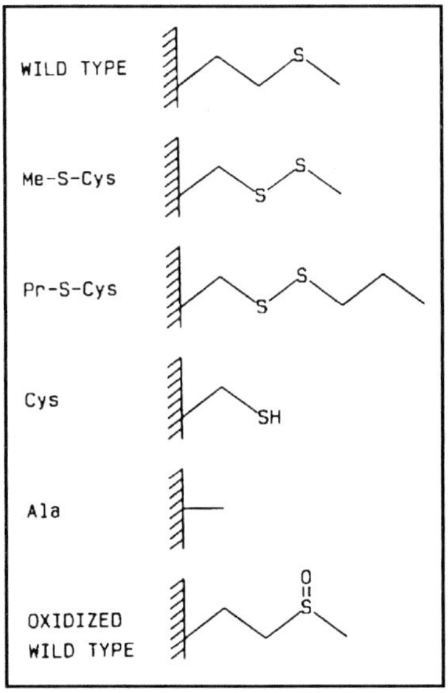

Fig. 8.4. Schematic representation of the structures of the side chains and side-chain derivatives occupying position 222 in subtilisin 309. Reprinted with permission from: Grøn H, Bech LM, Branner S et al. Eur J Biochem 1990; 194:900. © 1990 European Journal of Biochemistry.

residue S221 in the active site. This methionine undergoes oxidation to the sulfoxide form. Although M222 is not part of the catalytic triad, the bulky methionine sulfoxide interferes with the S221 which initiates peptide bond hydrolysis (see Fig. 8.4). In this way the enzyme becomes inactivated. Wells and colleagues attempted to overcome this drawback by substituting each of the other 19 natural amino acids at position 222. Mutant M222C had the highest specific activity: its k_{cat} was actually twice that of wild type subtilisin while its k_{cat}/K_m was 56% of wild type's. (Unfortunately, however, Cys is another amino acid that is very prone to oxidation.) The Ala and Ser substitution mutants were also reasonably active (k_{cat}/K_m values 14% and 11% of wild type respectively), moreso than residues such as Ile, Val or Leu which would be considered homologous with Met. The mutants containing Ala, Ser or Leu showed greater oxidative tolerance than the wild type and retained activity even after prolonged exposure to hydrogen peroxide concentrations up to 1M.[41,47,63] A similar M222A substitution greatly increased the resistance of both Savinase and Esperase to 0.04% peracetic acid at 20°C, enzyme activity being measured with sAAPF-pNA. In the absence of oxidizing agents, however, this same substitution was entirely neutral in Savinase™ and slightly destabilizing in Esperase™. A Savinase™ M222A, G195E double mutant encapsulated in coated granules had greatly-increased storage stability in a European-style (bleach-containing) detergent under accelerated test conditions at 50°C. Here, activity was assayed by the proteolytic degradation of dimethyl casein, followed by determination of the primary amino groups of the resulting peptide fragments with trinitrobenzene sulfonic acid (see chapter 6).[42]

Grøn and co-workers combined site-specific mutagenesis with chemical modification to achieve an oxidation-resistant form of subtilisin.[64] The enzyme in question came from *Bacillus lentus* and was termed subtilisin 309. Here also the Met 222 was replaced by Cys but the Cys was then thiomethylated by reaction with a thioalkylating reagent, methyl methane thiosulfonate (Me-MTS) to yield a thiomethylated Cys derivative, dubbed (Me-S-Cys^{222}) subtilisin (Fig. 8.4). This procedure markedly increased the enzyme's resistance to hydrogen peroxide with minimal loss in performance. Against the substrate sAAPF-pNA, (Me-S-Cys^{222}) subtilisin had a k_{cat}/K_m value 56% that of the wild type subtilisin 309, while in a dimethyl casein assay it displayed 89% of wild type 309 activity. These values were significantly better than those for the unmodified Cys 222 mutant of subtilisin 309. The modification was performed at an alkaline pH (9.5) at room temperature and was terminated by gel filtration. The substrates used to measure activity were again sAAPF-pNA and dimethyl casein. Oxidation protocols utilized 0.1 M or 1.0 M hydrogen peroxide. Modification with Me-MTS led to a 30% increase in activity. Modification of all Cys residues by Me-MTS was complete within 60 minutes, as demonstrated by subsequent reaction with Ellman's reagent (DTNB; see Table 6.1) and the failure of phenylmercuric chloride (reacts with free-SH groups) to reduce activity in the modified Cys222 enzyme. Kinetic experiments with sAAPF-pNA as substrate indicated that K_m values scarcely varied but k_{cat} values did show some interesting alterations. Wild type subtilisin 309 had a k_{cat} of 91.2 sec^{-1} while the corresponding value for its oxidized counterpart was only 6.1 sec^{-1}. The M222C mutant's k_{cat} was 43.2 sec^{-1} but this increased to 59.0 sec^{-1} following methylation with Me-MTS. Kinetic constants could not be determined for dimethyl casein but here again the wild type 309 enzyme performed best while (Me-S-Cys^{222}) subtilisin and the unmodified (Cys^{222}) subtilisin displayed 89% and 39% of wild type activity, respectively. (Me-S-Cys^{222}) was much more resistant to hydrogen peroxide than wild type 309 (or (Cys^{222})) subtilisin: half lives in 1 M H_2O_2 were 3-4 hours and < 10 minutes, respectively. Grøn and colleagues noted that sulfur-containing amino acids (Met and Cys) at position 222 seem to benefit activity, as these are the two derivatives with the highest activities. They speculate that the sulfur atom may participate in a hitherto unrecognized interaction with the transition state intermediate. Another Cys 222 mutant, chemically modified with propyl methane sulfonate, was also more resistant to oxidation by hydrogen peroxide but had poorer kinetic performance. The thiol alkylating reagents changed the Cys222 into a disulfide derivative which is less prone to oxidation than either the Met or Cys residues (Fig. 8.4). The notably good performance of (Me-S-Cys^{222}) subtilisin may be due to structural resemblance between the modified Cys side chain and the wild type methionine (Fig. 8.4).[64]

Scientists at Gist-brocades (the Netherlands; this particular division has since been acquired by Genencor) have described how a similar strategy led to improved oxidative tolerance in their detergent protease, Maxacal.™ Oxidizing agents (included as bleaching agents in most European detergents) lead to inactivation of Maxacal™, despite its containing no cysteines. Hydrogen peroxide (20 mM) inactivates 90% of Maxacal activity while the stronger sodium perborate/ tetraacetylethylenediamine combination ($NaBO_3$, 20mM/TAED, 10 mMgenerates peracetic acid) or diperoxydodecanoic acid (DPDDA, 10 mM) lead to complete inactivation. (The substrate sAAPF-pNA was used to estimate activities.)[65,66] Attention focussed on Maxacal's™ three Met residues. Each was replaced with amino acids less prone to oxidation but only one positional substitution (Met 216) led to stabilization. Mutants at position 216 completely resisted hydrogen peroxide but remained vulnerable to the other oxidizing agents, although their inactivation was much slower than that of the wild type Maxacal™. After oxidation for 1 hour, the M216Q enzyme retained > 90% activity in peroxide, approximately 60% in perborate/TAED and close to 35% in DPDDA. In each of these agents, wild type Maxacal's

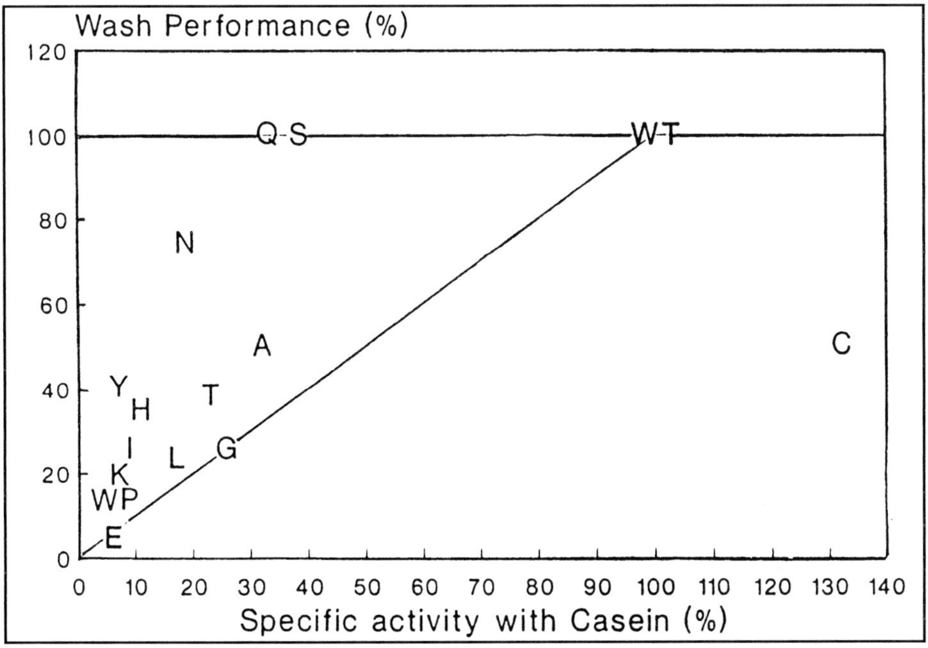

Fig. 8.5. Correlation between wash performance and specific caseinolytic activity of Maxacal™-M216 mutants. Letters signify individual amino acid replacements at position 216 according to the single letter amino acid code, except WT = wild type Met216. Reprinted with kind permission from: Misset O. Stability of industrial enzymes. In: van den Tweel WJJ, Harder A, Buitelaar RM, eds. Stability and Stabilization of Enzymes. Amsterdam: Elsevier, 1993: 124. © 1993 Elsevier Science - NL, Sara Burgerhartstraat 25, 1055 KV Amsterdam, The Netherlands.

activity had dropped below 50% within 10 minutes. The mutants M216S and M216Q each retained significantly greater activity than wild type over a period of weeks in a detergent storage test. Based on these findings, the researchers considered that peroxide damaged only the wild type Met216 while the stronger bleaches oxidize some other amino acid residue(s) leading to complete inactivation. Among the position 216 variants, activities on model substrates such as casein (or sAAPF-pNA) did not correlate at all with their actual wash performance characteristics (Fig. 8.5).[65,66] One of these oxidation-resistant variants has been commercially introduced as Maxapem.™

Subtilisin is potentially useful in organic synthesis (see, for instance, ref. 67) and has been tested in polar organic solvents, notably dimethylformamide (DMF). The effects of such media on enzyme stability have been discussed in chapter 3. Site-directed mutagenesis has been used in efforts to adapt subtilisin for solvent use. A form of subtilisin BPN' containing six individual site-specific mutations (M50F, G169A, N76D, Q206C, Y217K and N218S) showed remarkable stability improvements. Compared with the wild type, this mutant (dubbed 8350) was 100 times more stable at room temperature in aqueous solution and 50 times more stable in anhydrous DMF with a marginal improvement in kinetic characteristics.[68]

Back-mutation of position 217 to the wild type Tyr with retention of the other five mutations yielded derivative 8397. This variant had a half life of 350 hours in DMF compared with 20 minutes for the wild type while its pH 8.4 aqueous half life of 1600 hours dwarfed the wild type's 15 minutes under the same conditions.[69] The characteristics of the individual point mutations have been discussed by Bryan[40] who notes that the combination of individual stability-enhancing point mutations in a single subtilisin molecule allows one to achieve a high degree of stabilization in a step-by-step fashion. Similarly, von der Osten and colleagues report that the stabilizing increments in DSC (differential scanning calorimetry) temperatures of six individual Savinase™ mutations are nearly additive for the mutant incorporating all six changes (Savinase™ G195E, H120D, K235L, *36D, N76D and A194P). This six-substitution mutant proved to be 10.4°C more stable than the starting Savinase™, while the calculated stabilizing effect of the combined individual mutations was 11.2°C. In general, the effects observed on thermal denaturation temperatures correlated with storage stability data. Esperase, however, proved to be an exception: despite an increase of 2.2°C in its thermal denaturation temperature, the N76D Esperase™ variant was no more stable than its wild type counterpart during storage in a U.S.-style (no bleach) liquid detergent.[42]

Rational design of stabilized subtilisin mutants has been complemented by the use of random mutagenesis with phenotypic selection. Three individual mutants (M50F, I107V and K213R) each exhibited slightly greater alkaline stability than the wild type. The combination triple mutant was more stable than any of the single mutants.[70] (Note that the M50F mutation incorporated into the stable variants 8350[68] and 8397[69] had originally been discovered through random mutagenesis.) Other groups have also described stabilized mutants of subtilisin BPN' obtained via random mutagenesis[71] while Chen and Arnold have used the polymerase chain reaction (PCR) to create random mutants of subtilisin E which show improved catalytic properties in 85% dimethylformamide. A random double mutant D60N/Q103R with enhanced catalytic efficiency was stabilized by incorporation of a third mutation N218S which is known to enhance subtilisin's stability in polar organic solvents (see above). At 50°C in 40% dimethylformamide, the triple mutant had a half life of 9 hours compared with 6 hours for the wild type.[14] Arnold's group has extended this 'directed evolution' strategy to tailor the catalytic (not stability) characteristics of a para-nitrobenzyl esterase for optimal use in aqueous-organic solvents, especially in 30% dimethylformamide. Significant increases in the desired esterase activity accumulated from four sequential generations of random mutagenesis plus a single recombination step. Arnold notes that the actual, beneficial mutations could not have been predicted from the then-available structural homology model.[72] Halling argues that random mutagenesis represents a very good route for altering enzyme properties, given that design approaches still have notable limitations.[73] (See also 'Kanamycin nucleotidyltransferase' below).

Triose Phosphate Isomerase

Stabilization by protein engineering is not confined to monomeric proteins. Scientists at Massachusetts Institute of Technology (MIT) studied triose phosphate isomerase and whimsically abbreviated its name to TIM (MIT spelled backwards). The yeast enzyme is a dimer containing 12 asparagines per monomer, 3 of them at the surface of the interface between the constituent monomers. Catalytic activity depends on interactions between the monomers. Site-directed mutagenesis has successfully stabilized TIM. Replacement of the surface Asn78 by Ile or Thr yielded a 25% greater resistance to irreversible thermoinactivation at 100°C. The double mutant N14T, N78I was twice as stable as native TIM under these conditions. The Thr and Ile residues resist heat denaturation and approximate to the geometry of Asn.[52] When Asp replaced Asn78, the resulting mutant was 15% less stable at 100°C than the wild type, a finding that underlines Ahern and Klibanov's assertion[43] that deamidation of Asn residues to Asp is a key reaction in protein thermoinactivation. This N78D derivative was also noticeably less stable than the wild type under much less rigorous conditions (23°C, pH 7.9). Attempts to isolate a double Asn-to-Asp mutant were completely unsuccessful. Possibly, this was because such a derivative was so conformationally unstable that it suffered intracellular proteolysis during the bacterial fermentation.[52]

Case Study: Glucose Isomerase

Glucose isomerase (GI, also known as xylose isomerase) is used to prepare high fructose corn syrups (HFCS) for the soft drink industry. Usually immobilized, the enzyme converts glucose (derived from starch hydrolysis) to a mixture containing 42% fructose. Glucose syrup at a concentration of approximately 3 M is applied to a column reactor containing the immobilized isomerase. A typical temperature is 60°C. Inactivation of the enzyme during this process demands a lowering of the pump speed to maintain the conversion rate.[65] The industry, however, requires 55% fructose and an expensive chromatographic enrichment step must be applied to achieve this.[74,75] 55% Fructose can be achieved when GI is used in 85-90% ethanol at 30°C, but these conditions seem to labilize the enzyme.[74] Current glucose isomerases are microbial in origin and generally operate at temperatures in the range 60-65° or 70°C. Immobilization was originally undertaken to prevent waste of this costly enzyme but also improved operational stability by eliminating the need to include cobalt salts in the syrup.[75] There remains much scope for further stability improvement, however.

Loss of Glucose Isomerase Activity

An enzyme capable of operating at lower pH values (about 4.5) would prevent fructose by-product formation and would avoid the need to raise the pH of the saccharified starch feedstock entering the isomerization step.[75] Since the chemical equilibrium favours fructose over glucose at higher temperatures, a thermotolerant glucose isomerase could yield a 55% fructose product directly, without the need for enrichment.[74,75] Unfortunately, mere immobilization of GI is not sufficient to prevent inactivation at elevated temperatures, as evidenced in the following study. GI from *Streptomyces olivochromogenes* was purified and attached to aminated porous glass beads by glutaraldehyde coupling followed by borohydride reduction. At its pH optimum of 8.0, soluble GI lost activity when heated at temperatures in the range 80-90°C. This coincided with heavy aggregation which the presence of 1 M xylitol (a potent competitive inhibitor of GI[76] and widely used as a stabilizer for other proteins;[77] see chapter 5) could not prevent.[78] Immobilized GI was not subject to such aggregation, of course, but did undergo thermoinactivation at 80°C by an irreversible first order process with a half life of 120 minutes. This value is not much greater than the 76 minute half life of free GI under the same experimental conditions and shows that the stabilization due to GI immobilization is marginal. (The similar half life values for free and immobilized GI suggest that the aggregation of free GI may not be the cause of inactivation and that it may simply be an accompanying event.[78]) At 90°C the immobilized GI lost activity instantaneously in the absence of substrate but in the presence of its substrate, HFCS, it had a half life of 37 minutes. (Use of the competitive inhibitor xylitol at a concentration of 2 M extended the half life to 77 minutes.[78]) Misset and colleagues at Gist-brocades showed that inclusion of 0.25 M glucose increased the inactivation rate of *A. missouriensis* GI, with a concomitant decrease in the content of active tetramer and increases in the amounts of inactive dimer and monomer. They proposed that inclusion of glucose leads to dissociation of the active tetramer arising from glycation of lysine residues at the subunit interface.[65,79,80]

Volkin and Klibanov investigated the chemical changes leading to GI inactivation. Evolution of ammonia occurs during thermal inactivation of GI. The amount of ammonia produced and its kinetics of evolution suggested that deamidation of Asn and Gln residues was responsible for activity loss. This was subsequently proven experimentally. It appears that deamidation of only a few (Asn+Gln) residues is sufficient to cause inactivation and that unfolding of GI is required to expose critical amide residues to solvent. A longer term inactivation occurring at the lower but industrially relevant temperature of 70°C was assigned to oxidation of the enzyme's Cys residues.[78]

Effects of Metal Ions

Gaikwad and colleagues studied the GI from *Streptomyces sp.* NCIM 2730. Like other GIs, this enzyme is tetrameric (M_r 160 k) bt the dimer (M_r 80 k) is also active, unlike the monomer.[81] The metal ions Mg^{2+} and Co^{2+} had different effects with respect to activation of GI and its protection against inactivation by urea and guanidine hydrochloride. Maximal activity was attained only in the presence of both ions, in the combination 5 mM Mg^{2+}/1 mM Co^{2+}. Magnesium alone at 5 mM elicited 65% of maximum activity while 1 mM cobalt alone yielded only 38% maximal activity. This shows that magnesium is the more potent activator of GI activity. When GI was incubated under adverse conditions and retained activity was determined, these effects were reversed. Cobalt at 0.5 mM afforded complete protection against 5.6 M urea for 1 hour while metal-free GI and GI with 1 mM Mg^{2+} each lost 35% of initial activity. Metal-free GI exposed for 15 minutes to guanidine hydrochloride retained approximately 72% and 18% activity at guanidine concentrations of 1 and 2 M respectively. In the presence of either metal, GI retained more than 84% activity in 1M guanidine hydrochloride. At 2 M guanidine, 'Mg^{2+}GI' was 22% active while the figure for 'Co^{2+}GI' was 60%. None of the GI fractions showed any activity at 3 M guanidine. Twenty minute incubations at elevated temperatures showed that cobalt gave complete protection up to 80°C. A cobalt/magnesium mixture (0.5/1 mM) showed about 10% activity loss at 75°C. Metal-free and 'Mg^{2+}GI' each lost 20% activity at 50° and >30% at 80°C. SDS-PAGE analysis revealed that 'Co^{2+}GI' was predominantly dimeric while most of the GI was monomeric (inactive) in its absence. Cobalt, therefore, maintains GI's active dimeric structure and prevents its dissociation to the inactive monomer.[82]

Denaturation and Inactivation Studies

Ghatge and coworkers carried out further studies on the oligomeric nature and structure-function relationships of the NCIM 2730 GI enzyme.[83] They investigated the enzyme's behavior under differing denaturing conditions: 0.1% SDS causing mild denaturation, 2 M guanidine for reversible inactivation and heat, leading to irreversible inactivation. Techniques used were enzyme assay, gel filtration (to assess differences in quaternary structure), fluorescence and circular dichroism. SDS treatment caused the tetrameric GI to dissociate to an active dimer retaining

80% of initial activity. (Other workers had earlier demonstrated this active dimer form.) An inactive aggregate of M_r 200 k and an inactive monomer (M_r 40 k) also occurred in small amounts. The dimer and tetramer had similar fluorescence spectra, although differing in intensity. SDS-treated enzyme showed an altered far-UV CD spectrum, suggesting some perturbation of secondary, though not of tertiary, structure. Refolding of the enzyme by gel filtration under renaturing conditions led to some regain of the active tetramer, although most GI remained as the active dimer. (This was the first evidence that the tetramer-dimer conversion is reversible.) In guanidine hydrochloride, GI lost 55% activity at 0.3 M denaturant and was completely inactive at 1.8 M guanidine. When gel-filtered in 2 M denaturant, GI eluted as an inactive dimer. The fluorescence spectrum of this form showed a slight red shift and was less intense than that of native GI. Where the starting concentration of denaturant was 1.2 M or less, up to 90% of GI activity could be regained. The predominant form was the active dimer but this had a notably altered circular dichroic (CD) spectrum (although its fluorescence spectrum resembled that of the native active dimer). Following irreversible heat inactivation, the major GI form was a monomer with a blue-shifted and notably quenched fluorescence spectrum. The CD spectrum suggested that the protein's alpha-helical content was considerably decreased. These results suggested that, relative to the tetramer, the active dimer D has a perturbed secondary structure but has an intact tertiary structure supporting full catalytic activity. Note that fully active GI renatured from guanidine chloride had altered secondary structure and an unperturbed tertiary structure. It seems that an intact tertiary, rather than secondary, structure is essential for catalytic activity. In contrast, the inactive dimer D* has an unperturbed secondary structure but a distorted tertiary structure; in this respect it resembles the 'molten globule' stage in protein folding. These findings allowed Ghatge and colleagues to propose a model for GI inactivation,[83] which can help in the design of a protein engineering strategy:

$$T \leftrightarrow D \leftrightarrow D^* \rightarrow M$$

where T signifies the active tetramer and M is the inactive monomer; D and D* stand for the active and inactive dimer respectively, as defined above.

This understanding of the processes involved in GI activity loss allowed stabilization strategies to be devised. For the GI enzyme, protein engineering currently enjoys an advantage over the alternatives. While a number of thermophilic organisms produce apparently promising GIs (see chapter 4), most of these enzymes are not yet sufficiently characterized. Also, one must produce these enzymes from food-safe microorganisms for regulatory and consumer safety reasons. The most stable GIs isolated to date come from *Thermus aquaticus* and *Thermus thermophilus*. Both free and immobilized forms of the former enzyme retained full activity at 70°C for 30 days even in the absence of substrate. Unfortunately, however, the *Thermus* GIs require Mn^{2+} or Co^{2+} for activity and stability.[84] Neither ion is permitted as a food additive.[75]

Increasing the Stability of Glucose Isomerase by Mutation

Protein engineering has been successful in stabilizing GI. This is a considerable achievement since GI's molecular weight of 170,000 and its tetrameric structure of four identical TIM barrel subunits ('a dimer of dimers'[85]) make its protein engineering a very complex assignment. High resolution X-ray structures are available for GIs from three different bacterial genera.[75] The isomerase from *Actinoplanes missouriensis* was selected as the target protein. This enzyme inactivates irreversibly above 50°C at neutral pH but can be protected by Mg^{2+} or Co^{2+} ions up to 80-85°C. At 60°C, presence of glucose or other reducing sugars leads to faster inactivation than in their absence. This is thought to be due to a Maillard reaction between protein amino groups and reducing sugars. Wild type and mutant GIs were stressed at 85°C. Aliquots were removed periodically and stored on ice for later assay of catalytic assay. All GI forms lost activity by a first-order process and data fitted well to a single exponential curve. Lys253 lies at the enzyme's subunit interface. Lys253 and three other lysines (at positions 309, 319 and 323) were selected for mutation from molecular modelling studies. When Lys253 was replaced by Arg via site directed mutagenesis, the mutant GI (K253R) had a 4-6 fold higher thermostability at 60°C in the presence of 0.25M glucose and its kinetic parameters differed little from the wild type's. This may not seem particularly high but doubling the half life means doubling the productivity: this degree of stabilization could make a significant commercial difference. A likely explanation is that replacement of the Lys prevents a Maillard reaction between the lysine's epsilon-amino group and the reducing monosaccharides present in high concentration. This reaction is common in food chemistry and contributes greatly to GI inactivation under process conditions.[65,75,80,85] The glycated Lys residue produced by the Maillard reaction leads to destabilization of the tetrameric structure and to inactivation. Arg residues do not undergo this reaction to the same extent. At 84°C in the absence of glucose K253R was again more stable than the wild type, indicating that the Arg side chain can itself enhance GI stability by improved electrostatic interactions.[85] Replacement of lysine by glutamine in the indicated positions proved destabilizing at 84°C. Glutamine was used as a control for the loss of the positive charge associated with both Lys and Arg: note, however that it can undergo deamidation at elevated temperatures.[78] None of the other Arg substitutions yielded significant stabilizations over the wild type when glucose was included. Substitution of Arg for Lys is a very conservative mutation: both are positively charged. Positive charge appears to be very important in these locations since replacement by the electrically neutral glutamine destabilized GI. Arginine's higher pK_a renders it less reactive than lysine, however, and arginine is more likely to provide extended hydrogen bonding patterns due to its resonance stabilization. The Lys253→Arg substitution may also lead to additional hydrogen bonding and extra stabilization.[85]

In an interesting footnote to this investigation, the authors point out that the heat stable GI from *Thermus thermophilus* contains Arg at positions 253 and 309,[86] two of the positions where Arg replaced Lys in Mrabet et al's experiments,[85] showing that natural selection long ago came to the same conclusion as Mrabet and colleagues. One should recall here that Zuber (chapter 4) bserved a preference for Arg over Lys as lactate dehydrogenase enzymes from increasingly thermophilic *Bacillus* species were compared[24] and that, for a number of proteins, the Arg/Lys ratio is larger in thermophilic than in mesophilic examples.[23]

Attempts have also been made to influence the reversible unfolding (denaturation) that precedes inactivation of *A. missouriensis* GI. A triple mutant was constructed (Gly70→Ser, Ala73→Ser, Gly74→Thr) where certain small amino acid side chains were replaced by much bulkier ones so as to limit the conformational flexibility of the unfolded state and thus favor the folded state. In buffers at 85°C the mutant was more stable than the wild type but this was not mirrored under industrial conditions.[79] Here the extra stability was less than that gained by the Lys253→Arg mutation.

Copper, Zinc Superoxide Dismutase

The stabilizing Lys→Arg substitutions described above may be more generally applicable, as results with human copper, zinc superoxide dismutase (CuZnSOD) and *Bacillus subtilis* D-glyceraldehyde-3-phosphate dehydrogenase (GAPDH) show.[85] CuZnSODs are unusually stable homodimeric proteins[87] but human CuZnSOD is less stable than that from other species. Although human CuZnSOD shares the highly conserved Lys9 with other CuZnSODs, molecular modelling suggested that Arg at this position could have a stabilizing effect through the formation of several new hydrogen bonds. Such a K9R mutant was prepared by site-directed mutagenesis. It had enzymatic activity equivalent to the wild type but its 85°C half life was almost 6-fold greater than that of the wild type (98 and 17 minutes, respectively). GAPDH is a tetrameric enzyme and the thermostable form occurring in *Bacillus stearothermophilus* has Arg at position 281. This is replaced by Gly in the labile *Bacillus subtilis* GAPDH. Two mutant *B. subtilis* GAPDHs were constructed: one contained Arg at position 281 (G281R), the other Lys (G281K). Neither substitution affected enzyme activity but both increased thermal stability. Wild type *B. subtilis* GAPDH will not tolerate incubation at 75°C and has a 19-minute half life at 50°C. The mutants, in contrast, showed half lives at 75°C of 5 (G281K) or 18 minutes (G281R). These substitutions appear to give rise to improved intersubunit contacts. The gain in stability is comparable to the difference observed between proteins from mesophiles and thermophiles.[85]

Kanamycin Nucleotidyl Transferase

In the examples above, tertiary structural information was available either for the protein of interest or for a good homologue. Absence of three-dimensional structural information need not prevent isolation of stabilized enzyme derivatives, however: note that many stabilizing mutations of subtilisin arose from a program of random mutagenesis. In the case of kanamycin nucleotidyl transferase, Matsumura and Aiba used a biological thermal adaptation strategy which allows the selection of 'worthwhile' mutants. The mesophilic structural gene was mutated with hydroxylamine and then introduced into a vector plasmid. The plasmid was used to transform the thermophile *Bacillus stearothermophilus* and the resulting transformants were selected for increased thermal stability. The selection involved a temperature shift from 55°C to a non-permissive 61°C where the wild type enzyme loses activity. Two stabilizing single-residue substitutions—D80Y and T130K—were identified. The extra stability of both mutants was confirmed by in vitro experiments after purification.[88] Liao and colleagues reported similar findings only a short time later and also noted the D80Y change.[89] All of these mutants resisted irreversible thermoinactivation to a greater degree than did the wild type kanamycin nucleotidyl transferase.

An increase in the incubation temperature up to 70°C imposed further selection on the stabilized mutants. Mutants derived from this second round of selection were even more thermostable and turned out to have the second T130K mutation noted above. Both mutations occurred in regions predicted to have an alpha-helical structure. This isolation of identical mutations in different laboratories may be due to the presence of mutagenic 'hotspots' at these sites or, as noted in ref. 89, to a restricted number of possible changes that maintain activity while increasing thermostability.

To ascertain why the D80Y mutation stabilizes kanamycin nucleotidyl transferase, Matsumura et al undertook site directed mutagenesis of the wild type gene to replace Asp80 with residues other than tyrosine, namely Ser, Thr, Ala, Val, Leu,

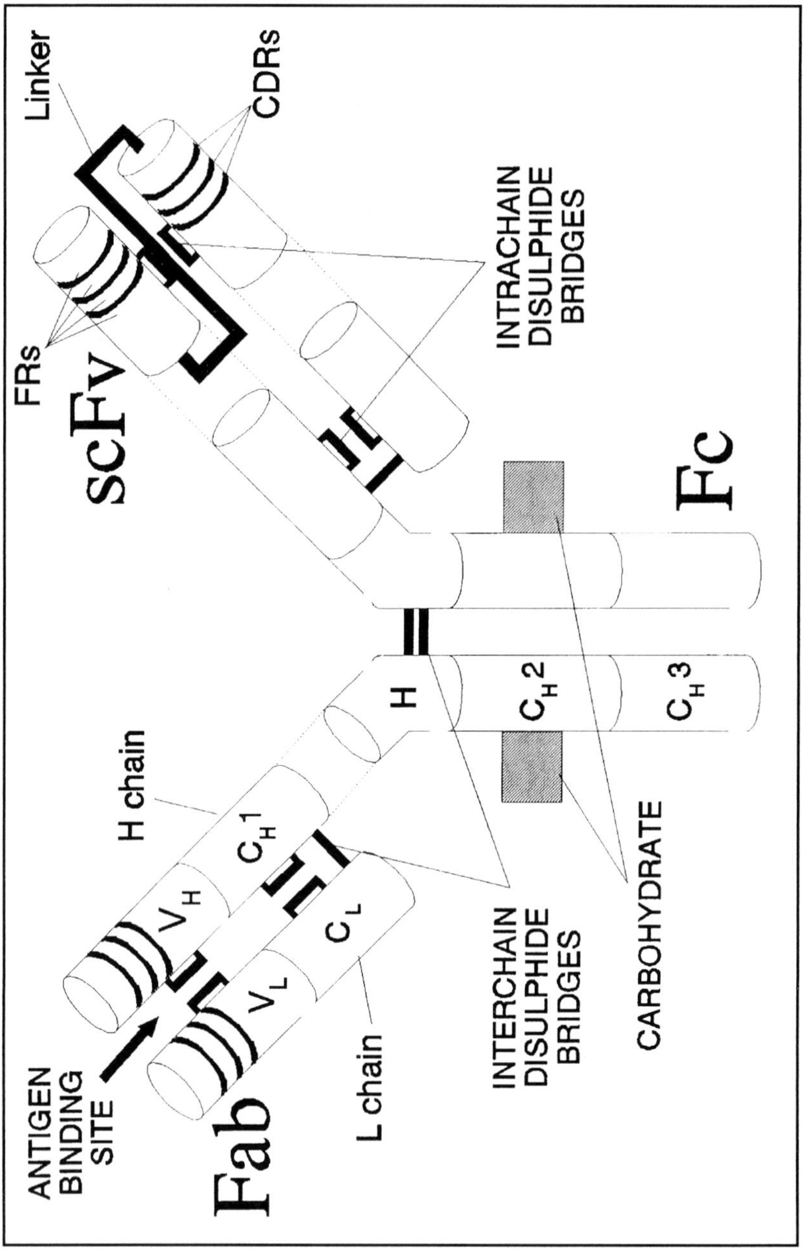

Fig. 8.6. Schematic diagram of the main features of antibody structure. © 1997 AJ Killard.

Table 8.1. Applications requiring stabilized antibodies

1. Use at high temperatures
   e.g., for immobilization protocols or for catalysis of certain reactions (catalytic antibodies).
2. Development of layers that remain active over long periods
   e.g., in sensor detection systems.
3. Use in vivo
   e.g., for imaging of diseases, for treatments such as drug/radionuclide targeting, for removal of toxic drugs. In these cases, resistance to i) proteolytic degradation, ii) too-rapid clearance from the circulation and iii) loss of binding capacity, is essential.
4. Use in in vitro immunization
   e.g., in treatment of certain diseases, including AIDS.
5. Generation of fragments for the production of bi- and multispecific antibody complexes
   e.g., for targeting toxins or cytotoxic cells to tumors for therapeutic purposes.

Phe and Trp.[90] Rate constants for heat inactivation at 58°C were measured. Half life values correlated with the hydrophobicity of the residue occupying position 80. The Tyr80 mutant proved to be more stable than its Phe80 counterpart, implying that a carefully-positioned hydroxyl group also plays a role in stabilizing the enzyme.

The thermal adaptation strategy is particularly suitable for the isolation of stabilized functional proteins. By selecting for catalytic (or other biological) activity, one can avoid the possibility of generating more stable, but functionally less efficient, mutants.[89,90]

Stabilization of Antibody Fragments

The use of protein engineering to stabilize proteins is not confined to enzymes. Antibodies play a number of important biotechnological roles as therapeutics,[91] diagnostics[92] and sensors.[93] The basic structure of a typical antibody (or immunoglobulin) is shown in Figure 8.6. It consists of two light (L) chains and two heavy (H) chains which are bound together by disulfide bonds. The heavy chains have molecular weights of 50-70 kDa whereas the light chains have weights of about 25 kDa. The heavy and light chains are made up of domains of about 160 amino acid residues; there are two domains per L chain and 3-5 per H chain, depending on the antibody class. Antibody chains consist of a chain of amino acid residues. At the N-terminal end of the light and heavy chains is the area which binds to the antigen, the molecule with which the antibody reacts. Both the heavy and light chains possess variable (V) and constant (C) regions. Within the variable regions, there are highly variable (hypervariable) sections termed the *complementarity determining regions* (CDRs) as well as other, relatively constant, segments. The constant regions provide a framework and, thus, a three-dimensional structure forming a pocket (the antigen binding site) on the antibody molecule. The design of this site allows the various CDR regions to interact directly with the antigens. Hence, a change in the amino acids comprising the CDR region can alter the antibody's binding specificity. Recent work has shown that the heavy chain variable region plays a crucial role in determining antibody specificity. It seems that the light chain stabilizes and enhances binding. This is probably very important under physiological conditions and may also be noteworthy when using antibodies for sensors.[93] It has also been suggested that amino acids in the variable region, but outside the CDR sites, may influence the correct placement of the antigen binding sites.

In many applications, however, the full antibody structure is not required (or desired) and it is common to generate defined antibody fragments by chemical or recombinant means.[93,94] Phage display techniques are especially useful for antibody generation.[95,96] This is facilitated by antibodies' well defined domain structure (Fig. 8.6). Usually it is the variable domains which are the most useful: these fragments (abbreviated to Fv) may bind to the target antigen with a lower affinity than the whole antibody molecule. However, the V_H and V_L chains comprising the variable domain fragment are much more liable to dissociate than are the corresponding regions of a full antibody.[97] (Dissociation is not a problem for Fab antibody fragments. These contain constant domains (C_H1 and C_L, which hold the variable domains together) and are much larger than Fv fragments.) Accordingly, there is a real need to stabilize Fv antibody fragments, in order to exploit their full potential in biotechnology. Some of the applications of stabilized antibodies (or antibody fragments) are listed in Table 8.1; the number of applications is increasing rapidly.

It is also possible to generate V_H and V_L domains, CDRs, antigen binding peptides and combinations of these.[98,99] In general, as the size of the unit diminishes, there is a decrease in binding stability. Often, this is associated with an increase in non-specific reactions. Evidence suggests that V_H domain and the heavy variable region CDRs may play a greater role in determining antibody specificity and stability. However, this may not always be the case. Since so much is now known about the structure of antibody binding sites, and since the technology to manipulate them is now quite well developed, work should focus on maximizing the stability of antigen-binding moieties, in terms of both their binding capacity and their persistence of activity in biological matrices (e.g. when used in vivo) or in synthetic arrays (e.g. in sensors). The stability of administered antibody fragments in vivo may differ markedly from that of an intact antibody, due to factors such as susceptibility to proteolysis, plasma clearance, tissue targetting and excretion.[97] Here we will focus only on stability in vitro, that is, on the thermodynamic and kinetic stabilities considered up to now.

Covalent crosslinking of the V_H and V_L chains is perhaps the most obvious way to prevent their dissociation. Three distinct types of crosslink have been explored. Two of these involved genetic alteration of the fragment (to insert a peptide linker and a disulfide bond) while the third was achieved by chemical modification. Fragments linked by any of these means were functional and did not undergo dissociation; they were also less prone to irreversible inactivation at 37°C.[100] The peptide linker joined the two chains to give a single chain Fv polypeptide (scFv), which could fold satisfactorily in *E. coli*.[101,102] The unfolding Gibbs energies of the V_H and V_L chains linked in either orientation (C-terminus of V_H to N-terminus of V_L, or vice versa) were virtually identical[103] and varied little with the linker's length.[104] NMR studies suggest that the linker is highly mobile.[105] Together, these results show that the peptide linker has little stabilizing effect beyond shifting the V_H-V_L dissociation equilibrium towards the associated state. The 37°C half life of the linked Fv fragment was ten times longer than that of the 'normal' Fv.[97] Disulfide bonds were inserted at positions L55-H108 and L56-H106 following a survey of candidate locations. Both variants could fold (and form the -S-S- link) in *E. coli* and their binding to the target antigen was almost identical to that of the Fv without either disulfide linker. The 37°C half lives of the linked fragments were 60 times longer, much greater than the degree of stabilization afforded by the peptide linkers.[97,100] Unfortunately, these disulfide bonds are not generally applicable: they actually link two of the CDR regions responsible for binding to the antigen and so these two -S-S-linked Fv fragments are not functional.[97] Later, suitable positions for interchain disulfide bonds were ascertained from superimposition of the Fv portions of ten known antibody crystal structures.[97] In most cases, two types of framework crosslinks were noted; these may be suitable for stabilizing antibodies of unknown three-dimensional structure.[97] Individual domains within an antibody molecule may undergo partial unfolding; this is likely to influence the molecule's overall stability characteristics.[97,106]

Ito and Kurosawa developed a system which allows display of Fv antibody fragments on a phage surface.[96] The V_H and V_L chains are associated only non-covalently in this system. Thus, while mutations can be introduced in the CDRs, the sequences within the framework regions must be left intact to ensure correct V_H-V_L association and proper Fv function. This same group made a library of 512 Fv fragments derived from a monoclonal antibody D1.3 directed against hen egg lysozyme.[107] Within this library, nine amino acids at positions in the CDRs of the heavy chain were replaced by similar residues. Three fragments (dubbed Wild-O, #2 and #11) were chosen for study of the thermal stability of the V_H-V_L association.[108] Circular dichroism (CD) and HPLC were the techniques used in this stability study. CD spectra were recorded over a range of increasing temperatures and HPLC runs were carried out at the particular temperature used for incubation. Given that CD spectra indicate a protein's content of secondary structure, the experimental results are more likely to reflect V_H and/ or V_L unfolding than the degree of V_H-V_L association. The Wild-O fragment gave a set of similar spectra at 218 nm at temperatures up to 50°C and a second set of spectra resembling each other above 70°C. Spectra obtained between 55-65°C showed considerable change and were quite different from each other. For the Wild-O fragment, a 50% change in CD characteristics occurred at 62°C. The corresponding values were 52° and 50°C for the #2 and #11 derivatives, respectively. Alone, the V_H and V_L fragments each denatured at 43°C. It seems that dissociation of Fv into V_H and V_L leads to the denaturation of both and that their association prevents thermal denaturation, i.e., association stabilizes individual fragments. HPLC analysis (which would mirror the degree of V_H-V_L association) indicated that 50% Fv existed at 48°C. For the #2 and #11 variants, 50% Fv occurred at 42° and 36°C, respectively. Thus the relationship between the three Fv fragments is consistent as measured by CD or HPLC. The different values obtained are probably due to concentration differences between samples used in the two techniques, but may reflect the presence of some intermediate state. One important conclusion from the comparison of the three closely-related fragments is that amino acid changes in the CDRs alone may influence thermal stability and the strength of the V_H-V_L association, even where the framework regions are left untouched. This could have implications for Fv fragment engineering.

There are no crystal structures available for the vast majority of antibody molecules and molecular modelling simulations may not have sufficient accuracy to determine suitable positions for disulfide bonds.[97] Thus, the scope for structure-based site-specific mutation of Fv fragments appears to be somewhat limited at present. Interestingly, analysis shows that many sizeable cavities occur within known antibody structures.[97] Chapter 2 described cavity-filling experiments which resulted in protein stabilization and emphasized the importance to stability of tight protein core packing. This would seem to be a worthwhile stabilization strategy for antibodies. Unfortunately, however, the 'antibody cavities' were not in any way conserved between different antibody molecules, so that a general cavity-filling approach could not realistically be undertaken unless one has access to the three-dimensional structure of the antibody in question.[97]

Miscellaneous Examples

There are many other examples of protein stabilization by protein engineering. Pakula and Sauer have investigated thermostability of the Cro protein from Lambda phage. Following their discovery that a Tyr26→Cys mutation increased Cro T_m by 11°C and its stability by 2.2 kcal.mol^{-1} (ref. 109), they investigated the role of the surface Tyr26 by replacing it with a variety of amino acids via cassette mutagenesis. None of the substitutions caused major alterations of the native secondary structure as assessed by CD measurements. Polar residues at position 26 led to significant increases in T_m for Cro (7.5°C increase for Gln, 10°C for His and 14.5°C for Asp) with positive changes in the value of the Gibbs energy of unfolding: +1.4, +1.9 and +2.7 kcal.mol^{-1} for Gln, His and Asp, respectively. Other residues did not enhance thermostability to such a degree but only Trp destabilized Cro with respect to the wild type Tyr26. The stabilization can be explained by relief of a reverse hydrophobic effect. A non-polar residue on the protein surface (Tyr26 in this case) may be more exposed to aqueous solvent in the native than in the denatured state. The resulting reverse hydrophobic effect would oppose correct folding due to an unfavorable Gibbs energy term. Substitution of such an exposed non-polar residue by polar side chains would lead to greater thermostability.[110]

Wigley et al have reported on a stabilized *B. stearothermophilus* lactate dehydrogenase mutant in which Asn replaces the conserved Ile250.[111] This residue lies in the apoenzyme active site in a water-accessible hydrophobic area. The coenzyme's nicotinamide ring binds to this area and covers it. Of course, Asn is much more hydrophilic than Ile, so this substitution results in a greatly-reduced hydrophobic area and increased thermostability. While Asn is a notably labile residue (see ref. 112 and chapter 2), the advantage of 'hydrophilizing' the water-accessible coenzyme binding site obviously outweighs any thermolabilizing effect of Asn substitution. This alteration of a hydrophobic residue to a hydrophilic counterpart resembles the surface hydrophilization of trypsin and chymotrypsin that Mozhaev and colleagues accomplished by chemical means.[113]

The potential benefits of a carefully-designed program of random mutagenesis coupled with suitable phenotypic screening have already been outlined in the discussion of subtilisin stabilization above. This strategy has the advantage that one can implement it in cases such as alpha-amylase, where the exact inactivating event(s) are unknown and where a three-dimensional structure is still awaited.[65] The starch processing industry makes extensive use of thermostable alpha-amylases to aid liquefaction of starch slurries at temperatures up to 110°C.[65] Random mutagenesis led to a thermostabilized alpha-amylase (H133Y) which also had improved stabilty characteristics in the starch liquefaction process.[65,114] The H133Y mutant's half life at pH 7 and 90.5°C in presence of 10 mM calcium was more than 2-fold greater than the wild type's. Improved H133Y half lives occurred with 1 mM calcium at 95°C over a range of pH values from 5.5 to 8 but were most noticeable at pH values up to 7.[65,114] (See also chapter 4.)

Martinez and Arnold undertook substitutions of charged surface residues of alpha-lytic protease, with the aim of enhancing its tolerance of the water-miscible solvent dimethylformamide (DMF; specific conditions 84% v/v DMF at 30°C). Alpha-lytic protease is a serine protease with 198 amino acid residues and three disulfide bonds. The charged residues selected for replacement do not participate in catalysis and were judged least likely to be involved in favorable salt bridges or hydrogen bonds. The four sites targeted were Arg45, Arg78, Glu125 and Arg129. In all, seven stabilized mutants were prepared involving two of these positions (45 and 78); six of these variants featured large hydrophobic amino acids (Ile, Leu, Phe and Tyr). The charge-reversing substitution R45E had a half-life six-times longer in 84% DMF while R78L's half life was four times longer. The half life of the double mutant R45E-R778L was 626 minutes compared with 23 minutes for wild type alpha-lytic protease. Only marginal effects were observed in aqueous buffer (60°C) for any of these favorable mutations.[115]

Subtilisin and alpha-lytic protease have not been the only bacterial proteolytic enzymes to receive attention from enterprising protein engineers. Toma et al[116] studied the neutral protease (NP) from *Bacillus subtilis* and achieved stabilization by grafting a calcium-binding loop from thermolysin onto the NP molecule. (This strategy was later used by Braxton and Wells on subtilisin BPN': see above and ref. 54.) They accomplished this by replacing the flexible surface loop between residues 188-194 inclusive on NP with a 10-residue calcium-binding segment from thermolysin. Thermolysin and NP are similar proteins with about 50% homology and the three-dimensional structure of thermolysin has been used to predict a model structure for NP. The two molecules are, however, notably different at loop regions due to sequence differences and to some insertions and deletions. The insertion was verified by partial digestion of the mutant NP with *Staphylococcus aureus* V8 protease and by atomic absorption spectroscopy, which showed a third calcium ion bound to the NP insertion mutant while only two calciums, as expected, bound to the wild type NP. This latter finding proved that the insertion still functioned as a calcium binding site following grafting onto its new 'home' molecule. Upon extensive dialysis at neutral pH and 4°C with <1 mM free calcium, the insertion mutant was less stable (as determined by residual catalytic activity) than the wild type. Gel electrophoresis indicated that the inactivation was due to autolysis; also, the calcium ion concentration had a critical influence on the extent of autolysis. At free calcium concentrations of 1 mM, stabilities of both enzymes were equivalent. However, when the free calcium concentration was raised to 100 mM, the insertion mutant NP was twice as resistant as the wild type to irreversible inactivation. (Kinetic stabilities were measured at 55°C.) Conformational stabilities were determined by circular dichroism at 62°C and these data also indicated a 2-fold increase in stability of the NP insertion mutant in presence of 100 mM calcium. This study demonstrated that replacement of a flexible, unstructured surface loop in a protein by a metal-binding loop could be a feasible stabilizing strategy.[116]

The luciferase enzyme from the firefly *Luciola cruciata* catalyzes a bioluminescent reaction which has been exploited as a very sensitive test for ATP levels in bacterial counting and other analytical applications. The enzyme protein comprises 548 amino acids and has a molecular weight of 60,024. Kajiyama and Nakano[21] undertook random mutagenesis of the cloned gene with the objective of generating thermostabilized mutants. Hydroxylamine (0.8 M) was used as mutagen of the gene-bearing plasmid at pH 6.0 and three mutants with enhanced pH and thermal stability (up to 8- or 10-fold at 50°C) were isolated. Each of these had a Thr→Ile

(ACA→ATA) substitution at position 217 and a catalytic activity 30% greater than the wild type's. (Two of the mutants also had silent mutations and the third had a H246T substitution.) To investigate this stabilizing T217I substitution further, Kajiyama and Nakano went on to carry out site-specific mutagenesis at position 217, substituting each of the remaining 18 natural amino acids. (Hydroxylamine leads only to GC-to-AT transition mutations and so hydroxylamine mutagenesis cannot be used to create all of the possible amino acid replacements.)[21] Analysis of this series of mutants indicated a correlation between thermostability and hydrophobicity of the substituting residue: the Leu, Val, Cys, Met and Phe mutants were all more stable than the wild type Thr. (Indeed, the T217L variant was more stable than the initially-isolated T217I mutant.)[21] Kajiyama and Nakano noted a similar pattern between their results and those of Yutani and colleagues[36] for the alpha subunit of tryptophan synthase. In this latter case also, enzyme stability (as measured by tolerance of denaturant) increased linearly with the hydrophobicity of the substituting residue.

Concluding Comments

Protein engineering has already afforded many examples of stabilized proteins. However, one can predict the likely effects of any given substitution only to a degree. This important limitation applies even to an extensively-characterized protein such as subtilisin. This is not to comment unfavourably on the predictive principles advanced by Menendez-Arias and Argos.[23] These are necessarily broad principles which, despite their usefulness for mutational design, cannot take into account the possibility of disruption of key interactions in the protein or the introduction of molecular strain. The strategy of biological thermal adaptation[88,89] seems particularly suited to the isolation of proteins with thermostabilized activities and has the advantage that X-ray crystal structures are not required. Despite the advent of molecular modelling and advances in protein structure determination, random mutagenesis with phenotypic screening will likely remain important in the design of stabilized proteins. Random mutagenesis is useful in cases where structural information is incomplete or absent. Selection of stable mutants can be accomplished quite simply and the mutation characterized. In this way only stabilizing mutations are examined: recall that some designed disulfide mutants of subtilisin yielded little or no gains in stability.[46] The first random mutagenesis experiments for subtilisin stabilization used classic chemical mutagens such as nitrous acid[40,71] The more sophisticated PCR used by Chen and Arnold allows one to restrict the random mutagenesis to a particular set of positions.[14] Stemmer has reported a variant of the PCR which he calls 'DNA shuffling'. This permits the rapid 'evolution' of a protein under laboratory conditions. The technique involves recombination between random fragments of selected mutant genes (obtained by DNase I treatment) followed by reassembly (without primers) and amplification (with primers) of the newly-recombined DNA variants using a conventional PCR protocol. Stemmer devised this protocol on the basis that oligonucleotide cassette mutagenesis and error-prone PCR are not combinatorial techniques and explore only a limited sequence space. Efficient sequence evolution is better achieved by iterative homologous recombination. The technique was applied to the molecular evolution of an efficient, cefotaxime-cleaving *E. coli* beta-lactamase, with cassette mutagenesis and error-prone PCR protocols being run in parallel. (Cefotaxime is a very potent antibiotic against *E. coli*, with a minimum inhibitory concentration of only 0.02 micrograms per ml.) The latter two techniques led to only a 16 fold increase in the minimum inhibitory concentration of cefotaxime. In sharp contrast, three cycles of DNA shuffling followed by two cycles of back crossing with wild type DNA (to eliminate any non-essential mutations; selection on successively increasing antibiotic concentrations took place after each cycle) led to a 32,000 fold increase in the minimum inhibitory concentration of cefotaxime with six amino acid changes from the wild type sequence.[117] This result dramatically indicates the potential of the DNA shuffling technique when combined with an appropriate selection system.

References

1. Fersht AR, Leatherbarrow RJ. Protein engineering. Protein Eng 1986; 1:7-16.
2. Knowles JR. Tinkering with enzymes: what are we learning? Science 1987; 236:1252-1258.
3. Bryan PN. Protein engineering. Biotechnol Adv 1987; 5:221-234.
4. Nosoh Y, Sekiguchi T. Protein engineering for thermostability. Trends Biotechnol 1990; 8:16-20.
5. Oxender DL, Fox CF, eds. Protein Engineering. NY: Alan R Liss, 1987.
6. Rees AR, Sternberg SJE, Wetzel, R, eds. Protein Engineering: a Practical Approach. Oxford: IRL Press, 1992.
7. Nosoh Y, Sekiguchi T. Protein Stability and Stabilization Through Protein Engineering. Chichester: Elis Horwood, 1991.
8. Benner SA. Expanding the genetic lexicon: incorporating non-standard amino acids into proteins by ribosome-based synthesis. Trends Biotechnol 1994; 12: 158-163.
9. Anthony-Cahill SJ, Griffith MC, Noren CJ et al. Site-specific mutagenesis with unnatural amino acids. Trends Biochem Sci 1989; 14:400-403.
10. Bain JD, Switzer C, Chamberlin AR et al. Ribosome-mediated incorporation of a non-standard amino acid into a peptide through expansion of the genetic code. Nature 1992; 356:537-539.
11. Mendel D, Ellman JA, Chang Z et al. Probing protein stability with unnatural amino acids. Science 1992; 256: 1798-1802.
12. Wells JA, Vasser M, Powers DB. Cassette mutagenesis: an efficient method for generation of multiple mutations at defined sites. Gene 1985; 34:315-323.
13. Hagen-Mann K. Application of the polymerase chain reaction to DNA engineering. In: El-Gewely MR, ed. Site-Directed Mutagenesis and Protein Engineering. Amsterdam: Elsevier, 1991:123-130.
14. Chen K, Arnold FH. Enzyme engineering for non-aqueous solvents: random mutagenesis to enhance activity of subtilisin E in polar organic media. Biotechnology 1991; 9:1073-1077.
15. Zebala J, Barany F. Mapping catalytically important regions in beta-lactamase using two codon insertion mutagenesis. In: El-Gewely MR, ed. Site-Directed Mutagenesis and Protein Engineering. Amsterdam: Elsevier, 1991:181-188.
16. Malcolm BA, Wilson KP, Matthews BW et al. Ancestral lysozymes reconstructed, neutrality tested and thermostability linked to hydrocarbon packing. Nature 1990; 345:86-89.
17. Matsumura M, Signor G, Matthews BW. Substantial increase of protein stability by multiple disulfide bonds. Nature 1989; 342:291-293.
18. Matsumura M, Becktel WJ, Levitt M et al. Stabilization of phage T4 lysozyme by engineered disulfide bonds. Proc Nat Acad Sci USA 1989; 86:6562-6566.
19. Wetzel R. Testing molecular tolerances. In: The Biochemist. London: Bulletin of the Biochemical Society, 1993; 15:11-15.
20. Das G, Hickey DR, McLendon D et al. Dramatic thermostabilization of of the yeast iso-1-cytochrome c by an asparagine/isoleucine replacement at position 57. Proc Nat Acad Sci USA 1989; 86: 496-499.

21. Kajiyama N, Nakano E. Thermostabilization of firefly luciferase by a single amino acid substitution at position 217. Biochemistry 1993; 32:13795-13799.
22. Imanaka T, Shibazaki M, Takagi M. A new way of enhancing the thermostability of proteases. Nature 1986; 324: 695-697.
23. Menendez-Arias L, Argos P. Engineering protein thermal stability. J Mol Biol 1989; 206:397-406.
24. Zuber H. Temperature adaptation of lactate dehydrogenase: structural, functional and genetic aspects. Biophys Chem 1988; 29:171-179.
25. Lee C, Levitt M. Accurate prediction of the stability and activity effects of site-directed mutagenesis on a protein core. Nature 1991; 352:448-451.
26. Alber T, Matthews BW. The use of X-ray crystallography to determine the relationship between the structure and stability of mutants of phage T4 lysozyme. In: Oxender DL, Fox CF, eds. Protein Engineering. NY: Alan R Liss, 1987: 289-297.
27. Perry LJ, Wetzel R. Disulfide bond engineered into T4 lysozyme: stabilization of the protein toward thermal inactivation. Science 1984; 226:555-557.
28. Wetzel R, Perry LJ, Baase WA et al. Disulfide bonds and thermal stability in T4 lysozyme. Proc Nat Acad Sci USA 1988; 85:401-405.
29. Matsumura M, Becktel WJ, Levitt M et al. Stabilization of phage T4 lysozyme by engineered disulfide bonds. Proc Nat Acad Sci USA 1989; 86:6562-6566.
30. Matthews BW, Nicholson H, Becktel WJ. Enhanced protein thermostability from site-directed mutations that decrease the entropy of unfolding. Proc Nat Acad Sci USA 1987; 84:6663-6667.
31. Suzuki Y. A general principle of increasing protein thermostability. Proc Japan Acad Sci 1989; 65 ser. B:146-148.
32. Nicholson H, Becktel WJ, Matthews BW. Enhanced protein thermostability from designed mutations that interact with alpha-helix dipoles. Nature 1988; 336: 651-656.
33. Matsumura M, Becktel WJ, Matthews BW. Hydrophobic stabilization of T4 lysozyme determined directly by multipoint substitutions of Ile3. Nature 1988; 334:406-410.
34. Zhang X, Baase WA, Shoichet BK et al. Enhancement of protein stability by the combination of point mutations in T4 lysozyme is additive. Protein Engng 1995; 8:1017-1022.
35. Kellis JT, Nyberg K, Sali D. et al. Contribution of hydrophobic interactions to protein stability. Nature 1988; 333: 784-786.
36. Yutani K, Ogasahara K, Tsujita T et al. Dependence of conformational stability on hydrophobicity of the amino acid residue in a series of variant proteins substituted at a unique position of tryptophan synthase alpha subunit. Proc Nat Acad Sci USA 1987; 84:4441-4444.
37. Tomizawa H, Yamada H, Imoto T. The mechanism of irreversible inactivation of lysozyme at pH 4 and 100°C. Biochemistry 1994; 33:13032-13037.
38. Tomizawa H, Yamada H, Hashimoto Y et al. Stabilization of lysozyme against irreversible inactivation by alterations of the Asp-Gly sequences. Protein Eng 1995; 8:1023-1028.
39. Stephenson RC, Clarke S. Succinimide formation from aspartyl and asparaginyl peptides as a model for the spontaneous degradation of proteins. J Biol Chem 1989; 264:6164-6170.
40. Bryan PN. Engineering dramatic increases in the stability of subtilisin. In: Ahern TJ, Manning MC, eds. Stability of Protein Pharmaceuticals, Part B: *In Vivo* Pathways for Degradation and Strategies for Protein Stabilization. NY: Plenum Press, 1992:147-181.
41. Wells JA, Estell DA. Subtilisin - an enzyme designed to be engineered. Trends Biochem Sci 1988; 13:291-297.

42. Von der Osten C, Branner S, Hastrup L et al. Protein engineering of subtilisins to improve stability in detergent formulations. J Biotechnol 1993; 28:55-68.
43. Ahern TJ, Klibanov AM. The mechanism of irreversible enzyme inactivation at 100°C. Science 1985; 228:1280-1284.
44. Zale SE, Klibanov AM. Why does ribonuclease irreversibly inactivate at high temperatures? Biochemistry 1986; 25: 5432-5444.
45. Volkin DB, Klibanov AM. Thermal destruction processes in proteins involving cystine residues. J Biol Chem 1987; 262:2945-2950.
46. Mitchinson C, Wells JA. Protein engineering of disulfide bonds in subtilisin BPN'. Biochemistry 1989; 28:4807-4815.
47. Wells JA, Powers DB, Bott RR et al. Protein engineering of subtilisin. In: Oxender DL, Fox CF, eds. Protein Engineering. NY: Alan R Liss, 1987:279-287.
48. Pantoliano MW, Ladner RC, Bryan PN et al. Protein engineering of subtilisin BPN': enhanced stabilization through the introduction of two cysteines to form a disulfide bond. Biochemistry 1987; 26: 2077-2082.
49. Takagi H, Takahashi T, Momose H et al. Enhancement of the thermostability of subtilisin E by introduction of a disulfide bond engineered on the basis of structural comparison with a thermophilic serine protease. J Biol Chem 1990; 265:6874-6878.
50. Narhi LO, Stabinsky Y, Levitt M et al. Enhanced stability of subtilisin by three point mutations. Biotechnol Appl Biochem 1991; 13:12-24.
51. Pantoliano MW, Whitlow M, Wood JF et al. The engineering of binding affinity at metal ion binding sites for the stabilization of proteins: Subtilisin as a test case. Biochemistry 1988; 27:8311-8317.
52. Ahern TJ, Casal JI, Petsko GA et al. Control of oligomeric enzyme thermostability by protein engineering. Proc Nat Acad Sci USA 1987; 84:675-679.
53. Pantoliano MW, Whitlow M, Wood JF et al. Large increases in general stability for subtilisin BPN' through incremental changes in the free energy of unfolding. Biochemistry 1989; 28:7205-7213.
54. Braxton S, Wells JA. Incorporation of a stabilizing Ca^{2+}-binding loop into subtilisin BPN'. Biochemistry 1992; 31:7796-7801.
55. Gros P, Kalk KH, Hol WGJ. Calcium binding to thermitase. J Biol Chem 1991; 266:2953-2961.
56. Strausberg SL, Alexander PA, Gallagher DT et al. Directed evolution of a subtilisin with calcium-independent stability. Bio/Technology 1995; 13:669-673.
57. Outtrup H, Boyce COL. Microbial proteinases and biotechnology. In: Fogarty WM, Kelly CT, eds. Microbial Enzymes and Biotechnology, 2nd edn. London: Elsevier Applied Science, 1990:227-254.
58. Betzel C, Klupsch S, Papendorf G et al. Crystal structure of the alkaline protease Savinase™ from *Bacillus lentus* at 1.4Å resolution. J Mol Biol 1992; 223:427-445.
59. Hastrup S, Branner S, Norris F et al. Mutated subtilisin genes. International patent application WO89/06279: 1989.
60. Morinaga Y, Franceschini T, Inouye S et al. Improvement of oligonucleotide-directed site-specific mutagenesis using double-stranded plasmid DNA. Bio/Technology 1984; 3:636-639.
61. Nelson RL, Long GL. A general method of site-specific mutagenesis using a modification of the *Thermus aquaticus* polymerase chain reaction. Anal Biochem 1989; 180:147-151.
62. Von der Osten C, Hedegaard L, Østergaard P at al. Stabilization of the detergent Savinase™ by proline substitution. In: van den Tweel, Harder A, Buitelaar RM, eds. Stability and Stabilization of Enzymes. Amsterdam: Elsevier, 1993: 133-144.
63. Estell DA, Graycar TP, Wells JA. Engineering an enzyme by site-directed mutagenesis to be resistant to chemical oxidation. J Biol Chem 1985; 260:6518-6521.

64. Grøn H, Bech LM, Branner S et al. A highly active and oxidation-resistant subtilisin-like enzyme produced by a combination of site-directed mutagenesis and chemical modification. Eur J Biochem 1990; 194:897-901.
65. Misset O. Stability of industrial enzymes. In: van den Tweel WJJ, Harder A, Buitelaar RM, eds. Stability and Stabilization of Enzymes. Amsterdam: Elsevier, 1993:111-131.
66. van Eekelen CAG, Mulleners LJSM, van der Laan JC et al. New proteolytic enzyme. European Patent Application 0328229: 1989.
67. Riva S, Chopineau J, Kieboom APG et al. Protease-catalyzed regioselective esterification of sugars and related compounds in anhydrous dimethylformamide. J Am Chem Soc 1988; 110: 584-589.
68. Wong CH, Chen ST, Hennen WJ et al. Enzymes in organic synthesis: use of subtilisin and a highly stable mutant derived from multiple site-specific mutations. J Am Chem Soc 1990; 112:945-953.
69. Zhong Z, Liu LJ-C, Dinterman LM et al. Engineering subtilisin for reaction in dimethylformamide. J Am Chem Soc 1991; 113:683-684.
70. Cunningham BC, Wells JA. Improvement in the alkaline stability of subtilisin using an efficient random mutagenesis and screening procedure. Protein Eng 1987; 1:319-325.
71. Rollence ML, Filipula D, Pantoliano MW et al. Engineering thermostability in subtiliisn BPN' by in vitro mutagenesis. CRC Crit Rev Biotechnol 1988; 8: 217-224.
72. Moore JC, Arnold FH. Directed evolution of a para-nitrobenzyl esterase for aqueous-organic solvents. Nature Biotechnol 1996; 14:458-467.
73. Halling PJ. Is random mutation more rational? Nature Biotechnol 1996; 14:435-436.
74. Visuri K, Klibanov AM. Enzymatic production of high fructose corn syrup (HFCS) containing 55% fructose in aqueous ethanol. Biotech Bioeng 1987; 30:917-920.
75. Quax WJ. Thermostable glucose isomerases. Trends Food Sci Technol 1993; 4:31-34.
76. Lee YH, Wankat PC, Emery AN. Purification of glucose isomerase by affinity chromatography. Biotechnol Bioeng 1976; 18:1639-1648.
77. Schein CH. Solubility as a function of protein structure and solvent components. Bio/Technology 1990; 8:308-317.
78. Volkin DB, Klibanov AM. Mechanism of thermoinactivation of immobilized glucose isomerase. Biotechnol Bioeng 1989; 33:1104-1111.
79. Luiten RGM, Quax WJ, Schuurhuizen PW et al. New mutant glucose isomerase enzymes. European Patent Application 0351029: 1990.
80. Quax WJ, Mrabet NT, Luiten RGM, Schuurhuizen PW et al. Enhancing the thermostability of glucose isomerase by protein engineering. Bio/Technology 1991; 9:738-742.
81. Gaikwad SM, Rao MB, Deshpande VV. Structure-function relationship of glucose/xylose isomerase from *Streptomyces*: evidence for the occurrence of an inactive dimer. Enzyme Microb Technol 1993: 15:155-157.
82. Gaikwad SM, Rao MB, Deshpande VV. D-glucose/xylose isomerase from *Streptomyces*: differential roles of magnesium and cobalt ions. Enzyme Microb Technol 1992; 14:317-320.
83. Ghatge MS, Phadatare SU, Bodhe AM et al. Unfolding and refolding of glucose/xylose isomerase from *Streptomyces sp.* NCIM 2730. Enzyme Microb Technol 1994; 16:323-327.
84. Lehrmacher A, Bisswanger H. Isolation and characterization of an extremely thermostable D-xylose isomerase from *Thermus aquaticus*. J Gen Microbiol 1990; 136:679-686.
85. Mrabet NT, Van den Broeck A, Van den Brande I et al. Arginine residues as stabilizing elements in proteins. Biochemistry 1992; 31:2239-2253.

86. Dekker K, Yamagata H, Sakaguchi K et al. Xylose (glucose) isomerase gene from the thermophile *Thermus thermophilus*: cloning, sequencing, and comparison with other thermostable xylose isomerases. J Bacteriol 1991; 173:3078-3083.
87. Lepock JR, Arnold LD, Torrie BH et al. Structural analyses of various Cu^{2+}, Zn^{2+}-superoxide dismutases by differential scanning calorimetry and Raman spectroscopy. Arch Biochem Biophys 1985; 241:243-251.
88. Matsumura M, Aiba S. Screening for thermostable mutant of kanamycin nucleotidyl transferase by the use of a transformation system for a thermophile, *Bacillus stearothermophilus*. J Biol Chem 1985; 260:15298-15303.
89. Liao H, McKenzie T, Hageman R. Isolation of a thermostable enzyme variant by cloning and selection in a thermophile. Proc Nat Acad Sci USA 1986; 83:576-580.
90. Matsumura M, Yahenda S, Yasumura S et al. Role of tyrosine-80 in the stability of kanamycin nucleotidyltransferase by site-directed mutagenesis. Eur J Biochem 1988; 171:715-720.
91. Hale G, Dyer MJS, Clark MR et al. Remission induction in non-Hodgkin lymphoma with reshaped human monoclonal antibody CAMPATH-1H. Lancet 1988; ii (8625):1394-1399.
92. Tijssen P. Practice and Theory of Immunoassays. In: Burdon RH, van Knippenberg PH, eds. Laboratory Techniques in Biochemistry and Molecular Biology 15. Amsterdam: Elsevier, 1985.
93. Killard AJ, Deasy B, O'Kennedy R et al. Antibodies: production, functions and applications in biosensors. Trends Anal Chem 1995; 14:257-266.
94. Milstein C, Winter G. Man-made antibodies. Nature 1991; 349:293-299.
95. Clackson T, Winter G. In vitro selection from protein and peptide libraries. Trends Biotechnol 1994; 12:173-184.
96. Ito W, Kurosawa Y. Development of an artificial antibody system with multiple valency using an Fv fragment fused to a fragment of Protein A. J Biol Chem 1993; 268:20668-20675.
97. Plückthun A. Stability of engineered antibody fragments. In: van den Tweel WJJ, Harder A, Buitelaar RM, eds. Stability and Stabilization of Enzymes. Amsterdam: Elsevier, 1993:81-90.
98. Mayforth RD. Designing Antibodies. NY: Academic Press, 1993.
99. Borrebaeck CAK, ed. Antibody Engineering: a Practical Guide. NY: WH Freeman, 1992.
100. Glockshuber R, Malia M, Pfitzinger I et al. A comparison of strategies to stabilize immunoglobulin Fv-fragments. Biochemistry 1990; 29:1362-1367.
101. Bird RE, Hardman KD, Jacobson JW et al. Single chain antigen-binding proteins. Science 1988; 242:423-426.
102. Huston JS, Levinson D, Mudgett-Hunter M et al. Protein engineering of antibody binding sites: recovery of specific activity in an anti-digoxin single-chain Fv analogue produced in *Eschericha coli*. Proc Nat Acad Sci USA 1988; 85:5879-5883.
103. Knappik A, Krebber C, Plückthun A. The effect of folding catalysts on the in vivo folding process of different antibody fragments expressed in *Escherichia coli*. Bio/Technology 1993; 11:77-83.
104. Pantoliano MW, Bird RE, Johnson S et al. Conformational stability, folding and ligand-binding affinity of single-chain Fv immunoglobulin fragments expressed in *Escherichia coli*. Biochemistry 1991; 30: 10117-10125.
105. Freund C, Guth B, Ross A et al. Characterization of the linker peptide of the single chain Fv fragment of an antibody by NMR spectroscopy. FEBS Lett 1993; 320:97
106. Tischenko VM, Zavyalov VP, Medgyesi GA et al. A thermodynamic study of cooperative structures in rabbit immunoglobulin G. Eur J Biochem 1982; 126: 517-521.

107. Ito W, Iba Y, Kurosawa Y. Effects of substitutions of closely related amino acids at the contact surface in an antigen-antibody complex on thermodynamic parameters. J Biol Chem 1993; 268: 16639-16647.
108. Yasui H, Ito W, Kurosawa Y. Effects of substitutions of amino acids on the thermal stability of the Fv fragments of antibodies. FEBS Lett 1994; 353: 143-146.
109. Pakula AA, Sauer RT. Amino acid substitutions that increase the thermal stability of the lambda Cro protein. Proteins Struct Funct Genet 1989; 5:202-210.
110. Pakula AA, Sauer RT. Reverse hydrophobic effects relieved by amino-acid substitutions at a protein surface. Nature 1990; 344:363-364.
111. Wigley DB, Clarke AR, Dunn CR et al. The engineering of a more thermally stable lactate dehydrogenase by reduction of the area of water-accessible hydrophobic surface. Biochim Biophys Acta 1987; 916:145-148.
112. Volkin DB, Staubli A, Langer R et al. Enzyme thermoinactivation in anhydrous organic solvents. Biotechnol Bioeng 1991; 37:843-853.
113. Mozhaev VV, Siksnis VA, Melik-Nubarov NS et al. Protein stabilization via hydrophilization. Covalent modification of trypsin and alpha-chymotrypsin. Eur J Biochem 1988; 173:147-154.
114. Quax WJ, Laroche Y, Vollebregt AWH et al. Recombinant mutant microbial alpha-amylase. World Patent Application WO91/00353: 1991.
115. Martinez P, Arnold FH. Surface charge substitutions increase the stability of alpha-lytic protease in organic solvents. J Am Chem Soc 1991; 113:6336-6337.
116. Toma S, Campagnoli S, Margarit I et al. Grafting of a calcium-binding loop of thermolysin to *Bacillus subtilis* neutral protease. Biochemistry 1991; 30:97-106.
117. Stemmer WPC. Rapid evolution of a protein in vitro by DNA shuffling. Nature 1994; 370:389-391.

CHAPTER 9

Long-Term Storage of Proteins

Introduction

The laboratory scientist or biotechnologist will often need to store an isolated or purified protein for varying lengths of time. If the protein is an object of study, it will take some time to ascertain its properties. If it is a commercial end product, or finds use as a tool in some procedure, it will likely be used in small quantities over an extended period. The protein must retain as much as possible of its original, post-purification, biological (or functional) activity throughout this time. The storage period or 'shelf life' can range from a few days to more than one year. The protein's *long-term* or *kinetic* stability, as defined in chapter 1, becomes critically important under these conditions. Shelf life can depend on the nature of the protein and on the storage conditions. This chapter differs somewhat from the others in this book. It places less emphasis on the literature of protein stabilization in favor of offering a number of practical recommendations to help maintain one's hard-won and/or valuable protein. These pages (which should be read in the light of discussion of stabilizing additives in chapter 5) outline a range of measures to prevent or lessen the inactivating events discussed in chapter 2. This chapter describes how one can prevent deterioration due to microbial contamination and proteolysis and the correct use of low temperatures for extended storage. It also considers drying and freeze drying as processes for long-term protein preservation. Of course, one cannot guarantee that any or all of these necessarily-broad recommendations will 'work' for a given protein, or that a particular stabilization factor will result from a procedure that seems to 'work'.

Apart from extremes of temperature and pH (which one will naturally avoid as conditions for routine or long-term storage), a variety of factors can lead to loss or deterioration of a protein's biological activity, as described in detail in chapter 2. These include proteolysis, aggregation and certain chemical reactions.[1] Proteolysis may arise from microbial contamination or be an intrinsic hazard in the case of proteolytic enzymes. A protein may lose an essential cofactor, or an oligomeric protein's subunits may dissociate from one another, resulting in each case in loss of activity.[1] Adsorption to surfaces may also lead to inactivation.[1,2] The purely chemical reactions are few and well defined (chapter 2). Elevated temperatures favor all of these destructive processes but it is important to note that aggregation and deleterious chemical reactions can occur at moderate temperatures also. Virtually complete aggregation of lyophilized bovine serum albumin occurred over 24 hours at 37°C following addition of just 3 microliters of physiological saline. The degree of aggregation was less, but still significant, at lower temperatures. The underlying

Stabilizing Protein Function, by Ciarán Ó Fágáin.
© Springer - Verlag and Landes Bioscience 1997.

cause was formation of an intermolecular disulfide bond by a thiol-disulfide exchange. Ovalbumin, glucose oxidase and beta-lactoglobulin underwent similar aggregation under the same conditions, but by a somewhat different mechanism.[3] These findings underline how important it is to ascertain correct storage conditions for the protein of interest. Simple reliance on a refrigerator to minimize activity losses may not suffice over an extended period. Another factor may work against the refrigerator: a few proteins are more stable at room temperature than in the refrigerator and are said to be *cold labile* (see chapter 1).

All containers used to store pure proteins should be of good quality and should tolerate temperatures as low as -20° or even -70°C if freezer storage is desired or necessary. A number of manufacturers (e.g. Sarstedt, Germany or Nunc, Denmark; there are many others) supply pre-sterilized screw cap plasticware with good mechanical and low-temperature properties. One should clean glassware exhaustively and sterilize it by dry heat; screw caps or rubber stoppers which cannot withstand dry heat can be autoclaved.

Prevention of Microbial Contamination

Microbial contamination can lead to significant losses of a pure protein by proteolysis. Contamination can be detected by standard microbiological plating techniques, but one should always aim to avoid it in the first place. Even if one can achieve successful elimination or removal of contaminating microorganisms, the protein of interest may already have lost at least some activity or may have deteriorated in ways difficult to detect. Addition of antimicrobial compounds such as sodium azide or thiomersal (sodium merthiolate, a mercury-containing compound) can prevent microbial growth. Add sodium azide to a final concentration of 0.1%(w/v) or thiomersal to a final concentration of 0.01%(w/v). (Both of these compounds are poisonous: handle them with care. They are totally unacceptable in any product for administration.) Azide will inactivate oxygen-binding proteins such as hemoglobin or peroxidase. Do not discard azide (compounds or solutions) down laboratory sinks: it can accumulate in old lead piping lead to the formation of potentially explosive compounds.

Where one desires sterility but must avoid use of the toxic compounds above, filtration offers a useful alternative. (One can use both strategies together, of course.) A filter of pore size 0.22 micrometers will exclude all bacteria; indeed, this method is used in industry to sterilize labile materials which cannot be autoclaved or irradiated. Disposable membrane or cartridge filters, of pore size 0.22 micrometers for sterile filtration, are available from companies such as Gelman (Ann Arbor, MI, USA; see Fig. 9.1) or Amicon Inc. (Beverly, MA, USA; Fig. 9.1) with or without a Luer lock attachment for extra secure attachment to a hand-held syringe. Most of these have very low protein-binding capacities. Typically, one draws the solution to be sterilized into a syringe and then removes the needle or tube. Next, one connects the filter to the syringe nozzle, ensuring it is firmly mounted. One then uncaps a suitable sterile storage container directly beneath the filter outlet (using standard aseptic manipulations to avoid contamination of container or cap) and depresses the syringe plunger to force the protein solution through the sterilizing filter into the container, recapping it immediately. One cannot 'flame' plastic containers in a Bunsen burner as part of aseptic technique: it is much better practice to perform filtration operations of this sort in a Class 2 laminar flow microbiological safety cabinet, the design of which prevents contamination of sample. It is not always possible to use a filter as fine as 0.22 micrometers directly. Some biological matrices, particularly sera, will not filter effectively through a 0.22 micrometer filter alone. One may need to prefilter the material initially through a coarser 0.45 micrometer filter to which the desired 0.22 micrometer filter is connected in series. Alternatively, one can accomplish the finer filtration as a separate operation. One can best filter larger volumes (hundreds of milliliters, or liters) using a stack of filters clamped in a special filtration unit. A filter as coarse as 1 micrometer may be used directly in contact with the solution of interest, the stack comprising progressively finer filters until the sterilizing 0.22 micrometer filter is encountered at the bottom of the stack. Filter manufacturers' technical representatives can provide specialist advice for individual cases.

Avoidance of Proteolysis

It can be difficult to remove proteases completely during purification of a target protein. Unless the object protein is completely pure (homogeneous), even tiny amounts of contaminating proteolytic enzymes can cause serious losses of activity during extended storage periods. The molecular diversity of proteases complicates the situation: there are exopeptidases, which remove amino acid residues from the N- or C-termini, and endopeptidases, which cleave internal peptide bonds within protein chains. In addition, there are four types of protease classified by their molecular reaction mechanisms: the serine, cysteine (or thiol), acid and metallo-proteases.[4] Use of EDTA in the concentration range 2-5 mM should complex the divalent metal ions essential for metalloprotease action. Pepstatin A is a potent but reversible inhibitor of acid proteases. It is used at concentrations of around 0.1 micromolar, as are similar protease inhibitors. The compound phenylmethylsulphonyl fluoride (PMSF) reacts irreversibly with the essential serine in the active site of serine proteases, inactivating them. It can also act on some thiol proteases. It is typically used at a final concentration of 0.5-1 mM, following dissolution in a solvent such as acetone (it is poorly soluble in water).[5] Before addition, of course, one must ensure that none of these compounds will adversely affect the protein of interest.

If the protein of interest is itself a proteolytic enzyme, use of protease inhibitors is not feasible. One may need to store such a protein in dried form or as a freeze-dried preparation (see paragraphs below). Alternatively, one can place it in a solution with a pH value far removed from the protease's optimum pH. Trypsin, for example, is most active at mildly alkaline pH values. Daily stock solutions of trypsin are often prepared in 1 mM HCl, where the very acid pH value renders the enzyme effectively incapable of catalysis. This helps prevent autolysis during the course of the experiment. The enzyme molecule does not inactivate under these conditions and is fully active on dilution into a suitable assay solution.[6] Many suppliers offer specific inhibitors of proteases or classes of protease. These inhibitors are often peptides or proteins, e.g., aprotinin, soybean trypsin inhibitor. A protease inhibitor cocktail is available in tablet form (trade name 'Complete') from Boehringer Mannheim, who state that the product effectively inhibits serine, cysteine and metallo-proteases during protein extractions from a variety of sources.

Extremely Dilute Solutions

Very dilute protein solutions are highly prone to inactivation. This is especially true of oligomeric proteins where dissociation of subunits can occur at low concentration. The individual polypeptide chains comprising the oligomer may lack activity alone and/or may denature with consequent loss of activity. Protein solutions of concentration less than 1-2 $mg.ml^{-1}$ should be concentrated as rapidly as possible[5] by ultrafiltration, or by reverse osmosis using solid sucrose or polyethylene

Fig. 9.1. Examples of commercial devices for sterile filtration and for concentration of small-volume solutions A. Centrifugal devices for sample concentration by ultrafiltration (NanoSpin Plus) or cell harvesting/removal of cell debris (Z-Spin Plus). B. SuporFlow® 100/200 and Supor DCF™ 0.1/0.2 micrometer sterilizing filters for biopharmaceutical processing and ultra-pure water production. Photographs courtesy of Gelman Sciences Inc., Ann Arbor, MI, USA. © 1995, 1996 Gelman Sciences Inc. C. Microcon® concentrator for concentration and purification of samples with volumes 50-500 microliters. D. Centriplus™ concentrator for concentration of dilute protein solutions (<1 mg.ml^{-1}); samples may be up to 15 ml volume. E. Stirred cells (series 8000) for sample concentration; cells may be used with a range of separate ultrafilters. F. ProFlux™ M12 sanitary bench-top filtration system for volumes up to 100 liters. Photographs courtesy of Amicon, Inc., Beverly, MA, USA. © 1995 Amicon, Inc.

Fig. 9.1A.

Fig. 9.1B.

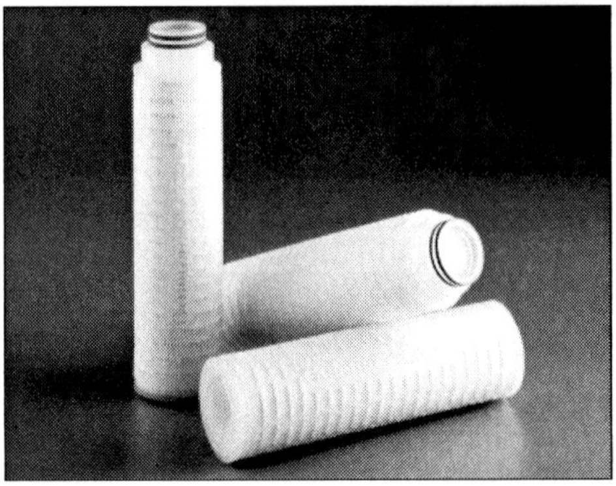

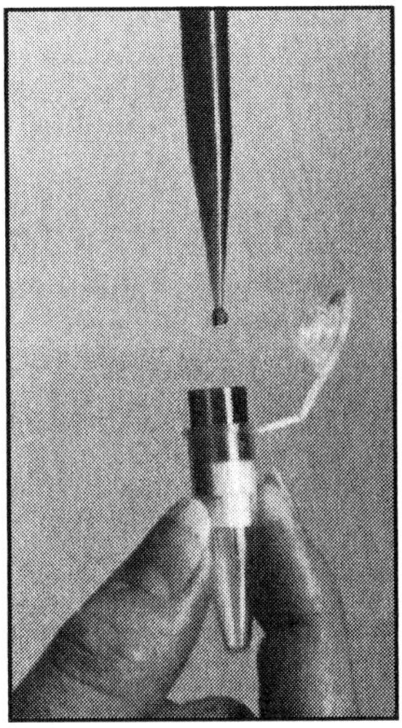

Fig. 9.1C.

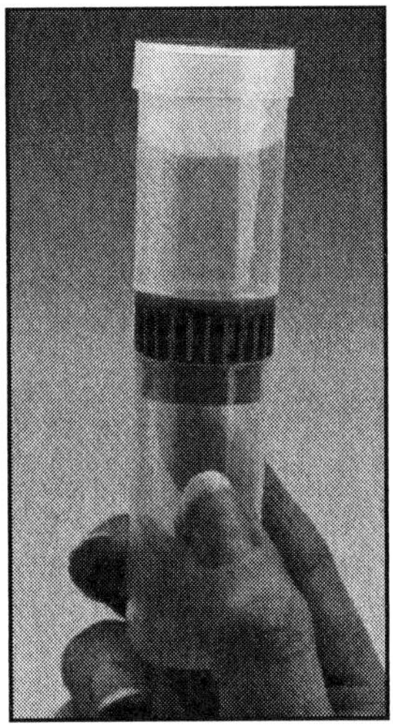

Fig. 9.1D.

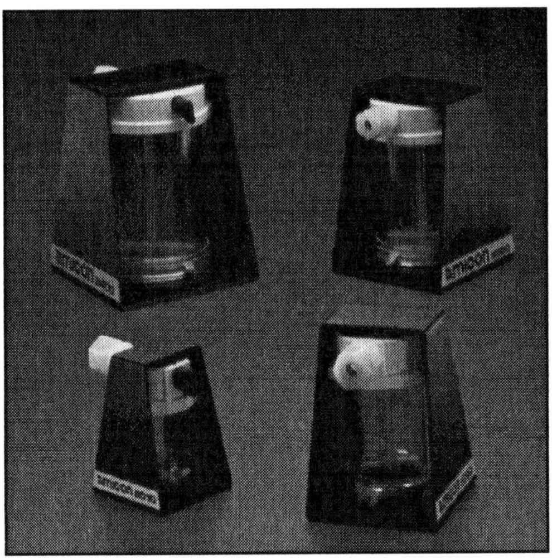

Fig. 9.1E.

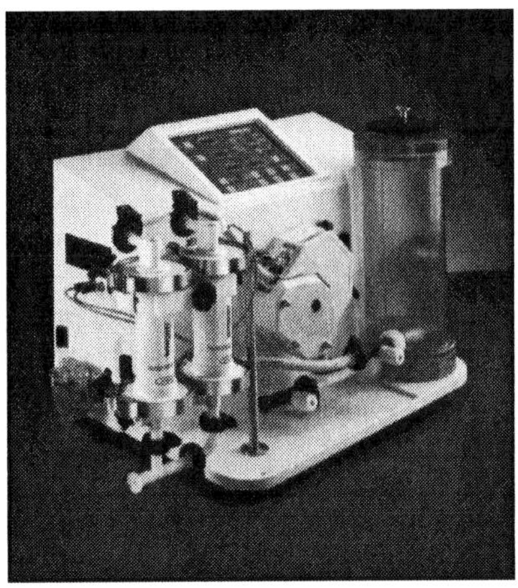

Fig. 9.1F.

glycol. A variety of vessels and membranes for laboratory-scale ultrafiltration, with a range of defined molecular weight cut-offs, are commercially available. These may comprise permanent stirred pressure cells with replaceable membranes (for volumes in the range 10-500 ml) or disposable centrifugal concentrators (for volumes up to 10 ml).[7] The Amicon and Millipore corporations are prominent suppliers of such apparatus. Figure 9.1 shows some typical examples. Schein gives some useful observations on ultrafiltration and suggests some other means of achieving protein concentration.[7] Where rapid concentration is not possible, one can prevent inactivation by addition of an exogenous protein such as bovine serum albumin (BSA), typically to a final concentration of $1 mg.ml^{-1}$. Alkaline phosphatase from *E. coli* is unstable at room temperature at concentrations less than 10 micrograms per milliliter but can be stabilized by addition of BSA.[8] Scopes has discussed possible reasons for the undoubted benefits of BSA addition.[5] BSA may act to prevent target protein adsorption to the container surfaces, it may prevent dissociation of subunits (in the case of an oligomeric protein) or it may even act in a chaperonin-like manner to prevent unfolding of the protein of interest.[5] To add an exogenous, contaminating protein such as BSA deliberately to a pure protein preparation may seem perverse, but occasionally this may be the price to pay in order to avoid inactivation.

Low Temperature Storage

Refrigeration at 4-6°C will often suffice to preserve a protein's biological activity, provided microbial contamination and proteolysis are prevented. Many proteins are supplied commercially in 50% glycerol or as slurries in approximately 3 M ammonium sulphate. Freezing of such preparations is unnecessary and should be avoided. Occasionally one may observe cold denaturation; this has been described briefly in chapter 1.

Use an ordinary household refrigerator for storage at temperatures of a few degrees Celsius. Modern machines often have a separate freezer chamber which can easily maintain temperatures as low as -20°C. Some proteins can deteriorate at 'refrigerator' temperatures and require storage at temperatures lower than 0°C. Usually, temperatures between -18° and -20°C, attainable by a household freezer, will allow for stable storage. Sometimes, however, it may be necessary to use temperatures below -20°C. In these cases, one needs to use a low-temperature laboratory freezer designed to maintain temperatures in the range -70° to -80°C. (These latter temperatures are extremely cold and can inflict a 'cold burn' on exposed skin. One should always wear insulating or autoclave gloves when handling low-temperature freezer items.)

Most protein solutions will freeze solid at temperatures below 0°C. (Mixtures containing high concentrations of glycerol, however, will remain liquid at -20°C: see chapter 5 and paragraphs on 'Freezing' below.) The events occurring during freezing of a protein-containing mixture or biological system are much more complex than the simple macroscopic phase change suggests. Differential freezing of particular components of the mixture can lead to enormous concentration effects and to dramatic changes of pH at low temperatures. These chemical processes can lead, in turn, to a notable degree of protein inactivation. The subject of freezing damage and its avoidance is discussed later ('Freezing' below). Note that one can often minimize the problem by judicious choice of stabilizing additives as discussed in chapter 5; see also ref. 9.

Prevention of freezing will, of course, avoid freezing damage. It is possible to undercool liquids without freezing by preventing the nucleation of ice crystals.

This means that proteins can be stored well below 0°C in the liquid phase. The preparation of protein-containing aqueous-organic emulsions which can maintain complete biological activity in the liquid state over extended periods at -20°C has been described[10] and special fluids for the preparation of undercooling emulsions are commercially available under the trade name 'U-COOL'. This method is very useful for small volumes of valuable proteins. It avoids the need to use additives and is more economical than freeze drying. The same process is used for many different proteins and one can remove portions of a sample without effect on the activity of the remainder. The actual storage temperature matters little, provided the upper temperature is less than 4°C and the lower temperature remains above -40°C, the nucleation temperature for ice crystal formation.[11]

Drying for Stable Storage

The advantages of water removal as a protein storage/stabilization strategy are many. Water participates directly in many of the deleterious chemical reactions discussed in chapter 2 and in proteolytic processes. In any case, it provides a medium for molecular movement and interactions. For these reasons, removal of water effectively prevents deterioration of the protein. A dried preparation will be much less bulky than the original solution and can conveniently be stored in a laboratory freezer or refrigerator (or perhaps even at room temperature). When one wishes to use the protein preparation, one can rehydrate it simply by addition of an appropriate volume of pure water or a suitable buffer solution.

At low temperatures, a liquid may behave in one of two ways. *Eutectic* solutions undergo a sharp liquid-solid freezing transition over a very narrow temperature range while *amorphous* liquids show a *glass transition* in which viscosity increases dramatically with cooling and the solution takes on the macroscopic properties of a solid, even though it does not crystallize. Below this glass transition temperature (T_g') virtually no adverse chemical or biological reaction can take place. (Curiously, however, enzyme activity has been shown to occur below the glass transition temperature[12] but the significance of these very low catalytic-center activities remains unclear.[13]) Above T_g', however, the viscous, rubbery material is very prone to deterioration[14]—see Figure 9.2. Understanding these low-temperature features of liquids is important for effective drying (and freeze drying, below).

It is possible to design protein-compatible formulations with glass transition temperatures (T_g') typically as high as 37°C.[15] With these high T_g' values, controlled evaporative drying can be used in place of lyophilization (freeze drying) to stabilize proteins in the solid state. Worthwhile evaporation rates will occur below these high T_g' values at reduced pressure. Evaporation is faster, less costly and more easily controlled than freeze drying.[11,15] The high T_g' values also mean that one can sometimes store the resulting dried product at ambient temperature: if room temperature is less than the glass transition temperature, the protein formulation will not undergo a glass/rubber transition during storage at room temperature. The glass-forming compounds are typically carbohydrates; maltose and maltohexose are particularly valuable while sucrose can be useful if the moisture content can be reduced to 2% or less.[15] One can reconstitute the solid protein preparation by simple rehydration with added water or buffer. This method has been patented and is described in ref. 16. An alternative formulation for vacuum drying of proteins, involving the use of a cationic, soluble polymer (e.g. diethylaminoethyl dextran) and the sugar alcohol, lactitol, has also been patented[17] and published.[18]

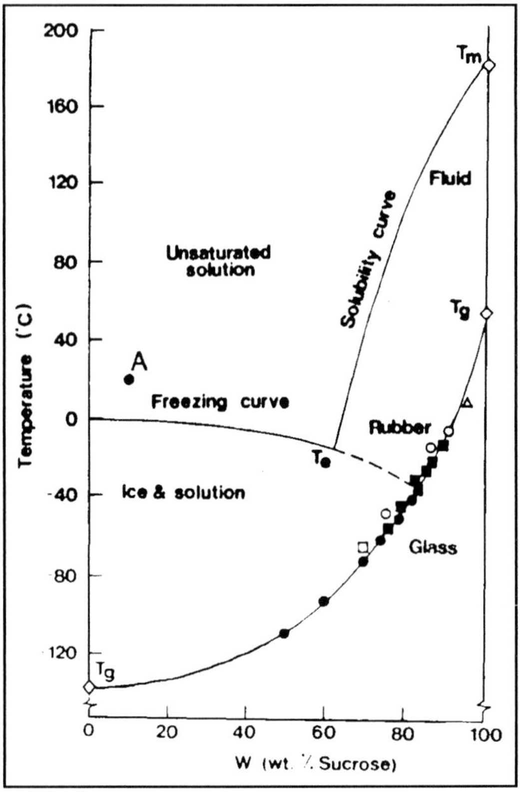

Fig. 9.2. Solid-liquid state diagram for the sucrose-water system showing the glass transition T_g/concentration profile, the equilibrium solubility curve, running between the melting point and the eutectic temperature, T_e, and the region of metastability (supersaturation) beyond T_e (broken line). Symbols represent experimental data taken from different sources. Point A represents a typical dilute solution that is to be dried to a stable solid product. Reprinted with permission from: Franks F, ed. Protein Biotechnology. Totowa, NJ: Humana Press Inc, 1993: 512. © 1993 Humana Press Inc.

Freeze Drying

Lyophilization, or freeze drying, is a method for the preservation of labile materials in a dehydrated form. It can be particularly suitable for high-value biomolecules such as proteins. The process involves the removal of bulk water from a frozen protein solution by sublimation under vacuum with gentle heating (primary drying). This is followed by controlled heating to more elevated temperatures for removal of the remaining 'bound' water from the protein preparation (secondary drying). Residual moisture levels are often lower than 1%.[19] If the freeze drying operation is carried out correctly, the protein will preserve all or most of its initial biological activity in the dry state. This dry state offers many advantages for long-term storage of the protein in question. Understanding of freeze drying as a process has grown in recent years[14,20] and knowledge of the effects of additives has also grown.[9,21] Together, these advances allow one to use lyophilization on a more rational and less strictly-empirical basis. Reference 15 contains a useful treatment which touches on the underlying theory.

Effective freeze drying takes time and the process cannot be hurried. Run times of 72 hours or greater are not unusual, depending on the nature of the product formulation and the properties of its constituents. A freeze dryer (lyophilizer) is a complex apparatus and it will have high capital and running costs (the latter due to its power consumption and the time required). Freeze drying is usually reserved, therefore, for high-value proteins or is used in cases where alternative product

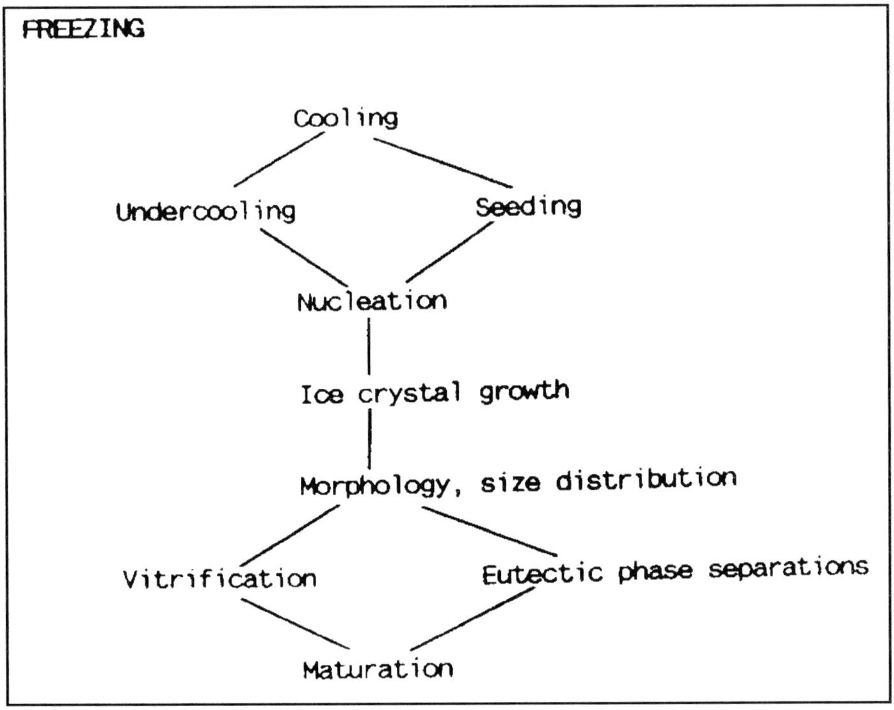

Fig. 9.3. Physical principles of freeze drying. Reprinted with permission from: Franks F, ed. Protein Biotechnology. Totowa, NJ: Humana Press Inc, 1993: 515. © 1993 Humana Press Inc.

presentations (such as ammonium sulfate precipitates) are unsuitable or give insufficient shelf-lives.

A typical freeze dryer comprises a *set of shelves* which can be cooled to about -40°C or heated above ambient temperature, a *condenser* usually cooled to -60°C and a high performance *vacuum pump* (see Fig. 9.4). The shelves allow rapid freezing of the protein solution and, later, gentle programmed heating for primary and secondary drying. The vacuum pump reduces the pressure within the system to allow sublimation of the bulk water (i.e. primary drying) while the condenser acts as a lower-temperature trap for the sublimed water which collects in the form of ice (i.e. it provides a temperature gradient within the apparatus). This explains why the condenser temperature must be considerably less than that of the shelves.

A complex interplay of chemical and physical phenomena takes place during freeze drying. The product yield (that is, the percentage recovery of the initial active protein) depends on the formulation in which the protein is placed before lyophilization[14,15,21] while its ease of rehydration and its stability on long-term storage (or 'shelf life') are influenced by the processing regime.[14] Some critical factors concerning the operations of freezing, primary and secondary drying are outlined in the following paragraphs. The final sections of this chapter describe, in broad terms, the operation of a typical freeze dryer apparatus.

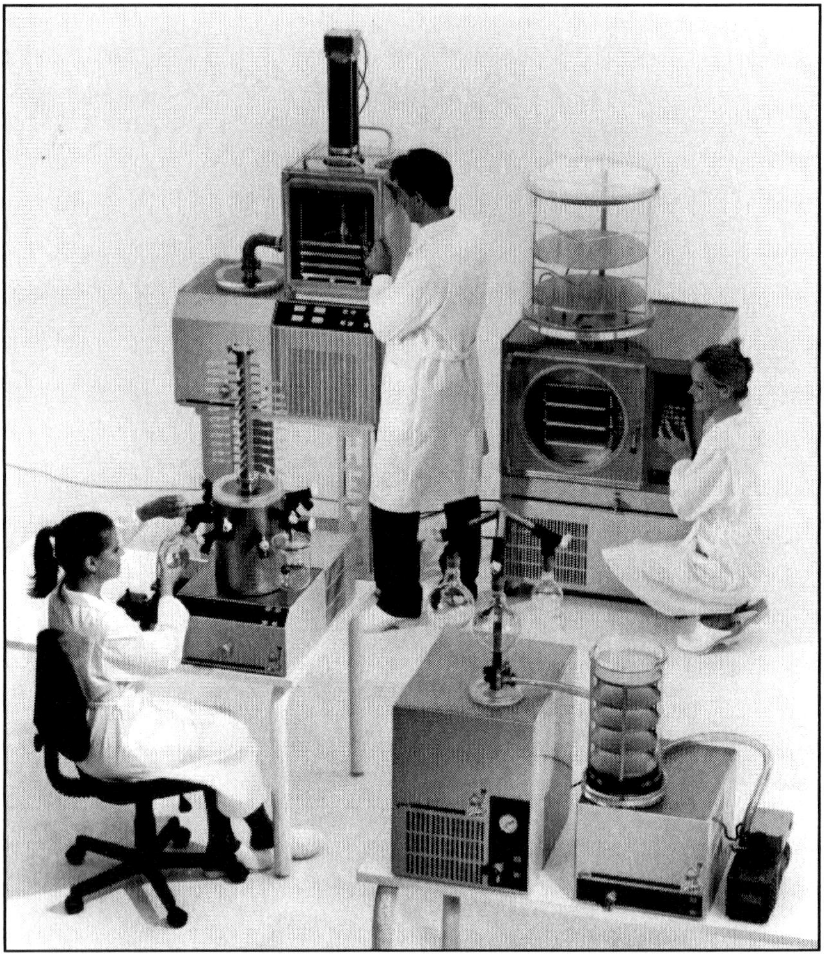

Fig. 9.4. Examples of commercial freeze dryers (A, this page; B, next page). A. Some different types of laboratory-scale lyophilizers. The two machines at the front feature manifold attachments for freeze drying the contents of individual flasks. Note the smaller manifold apparatus on top of the machine in the left foreground; this allows the processing of individual small ampules. The two larger machines in the rear of the photograph have temperature-controllable shelves inside a freeze-drying chamber. The position of the shelves in the freeze dryer at left rear can be lowered by use of the compressing piston that is visible in the photograph; this feature permits the full closure of loosely-stoppered vials of lyophilized product while still under vacuum.

Freezing

In freezing, liquid water crystallizes to yield solid ice. Water, in common with other substances, including many solutes, is a *eutectic* material with a sharp transition between the liquid and solid states. Non-crystalline materials undergo a *glass transition* and become rigid at a certain low temperature (T_g'). This is the end result of a notable increase in viscosity with decreasing temperature. The material becomes increasingly rubbery until rigidification occurs at T_g' and liquid movement effectively ceases. (Evidence suggests that protein molecules also un-

Fig. 9.4B. View of the drying chamber of a larger lyophilizer. Note the temperature-controllable shelves inside the chamber and the inspection ports set into, and projecting from, the open door. Note also the recording chart roll in the middle of the panel on the left-hand side, underneath the digital control pad. Photographs kindly supplied by Heto A/S, Allerød, DK-3450, Denmark. © 1997 Heto A/S.

dergo a loss of mobility at low temperatures and become completely rigid or 'frozen' at or below 220K [-53°C].)[22] Even for eutectics, ice formation usually does not occur at the thermodynamic freezing point: the actual temperature of freezing is generally 10-15°C lower than this so that supercooling (or undercooling) is required.[14,20] It is generally satisfactory to cool the product to just below 0°C initially, but not so low as to induce crystallization. When all the product has cooled to this set temperature, one can reduce the shelf temperature sharply to crystallize bulk water to ice. Typical temperatures lie in the range -20 to -30°C. One should allow time for ice to form in all containers before proceeding to the next stage, which involves further shelf cooling below the lowest eutectic temperature (for crystalline materials) or below the glass transition temperature (T_g', for amorphous substances). A usual final temperature is -40°C. The product is reliably solidified only below these temperatures and only now may primary drying begin. It is vital to maintain the freezing temperature below T_g' before and during primary drying. If the temperature of the material rises above this value, the material ceases to be a solid and becomes a viscous rubber very prone to deleterious reactions, with resulting losses of activity. The rubber will also undergo a mechanical *collapse*, take on a different appearance and be difficult to rehydrate.[14] Collapsed product is always unacceptable even if reasonable biological activity remains. One can determine the glass transition temperature by differential scanning calorimetry, although a sensitive instrument is needed.[14] Electrical resistance measurements can sometimes be useful but microscopic observation of freeze drying over a range of temperatures is arguably the most direct, sensitive and unambiguous method for determination of the collapse temperature.[20]

Other notable changes occur in the product during freezing, many of which can lead to significant protein inactivation. As the bulk water freezes to ice, the amount of liquid water remaining naturally decreases. This leads to great increases in the concentrations of solutes such as salts. Such freeze-concentration can have far-reaching effects. Concentration will increase the rates of unwanted chemical reactions such as oxidations. Buffer components may crystallize differentially, leading to pronounced pH shifts; also, pK_a values are temperature-dependent. For phosphate buffers, one should use potassium or mixed salts in preference to sodium salts.[14,23] For all of these reasons, one should accomplish the freezing steps (crystallization of bulk water and cooling below the eutectic or T_g' temperatures) in any lyophilization process as quickly as possible.

Suitable excipients or protective substances can lessen or overcome some of these damaging freezing effects: they can particularly influence melting or collapse temperatures.[24] Excipients may be classed as bulking agents, tonicity modifiers, buffers and cryoprotectants/lyoprotectants.[24] One should use bulking agents where the product's solids content is low. They help to ensure the development of a plug of dried material which will be of good appearance. Bulking agents can also prevent blowout: where the material has a very low solids content of the order of 1%, disruption of cake structure by the issuing water vapor can result in loss of the dried product from the container along with the vapor.[21] Typical bulking agents include polyols and certain sugars, mostly non-reducing. One must choose buffers with great care. Possible variations in pK_a and solubility with temperature, as well as compatibility with the protein(s) in question and other constituents, must be considered. Reference 25 contains useful information on buffers. Many bulking agents will double as tonicity modifiers in preparations for administration *in vivo* to ensure that the preparation is isotonic with body fluids. Any agent used for this purpose must, of course, be pharmaceutically acceptable. Cryoprotectants and lyoprotectants are substances which stabilize a protein against the deleterious effects of freezing and of lyophilization/storage respectively. These protecting additives appear to be preferentially excluded from the protein surface. This results in a strengthening of the water 'shell' surrounding the protein;[26] see chapter 5 for a more complete discussion. Polyols, certain salts and sugars can each be beneficial. In general, non-reducing sugars are preferable since they cannot participate in Maillard reactions. Xylitol's T_g' is -46.5°C in a freeze concentrate of 42.9 weight % water while T_g' values for sorbitol, sucrose and trehalose are -43.5°, -32° and -29.5°C at water weight % values of 18.7, 35.9 and 16.7, respectively.[14] Glycerol has a notably low T_g' value of -65°C at 46 weight % water.[14] This means that glycerol-containing formulations must reach very low temperatures to become glassy. Glycerol, therefore, is not an ideal lyophilization excipient, as such temperatures may be difficult to achieve. Useful discussions and examples of cryoprotectants occur in references 7, 9 and 26.

Primary Drying

Primary drying is the sublimation under vacuum of bulk ice from the product to the much colder condenser typically held at -60°C. Sublimation will be faster at higher shelf temperatures, so heat the shelves to a few degrees below T_g' to quicken the process. NEVER allow the shelf temperature to equal or exceed T_g', however: if it does, collapse and product deterioration may occur. A usual safety margin is 2-5°C below the eutectic or collapse temperature.[20] (The term 'collapse temperature' is usually equivalent to T_g' for an amorphous substance.) The rate of sublimation depends on the vacuum and on the condenser temperature and so is largely determined by the characteristics of the lyophilizer apparatus.[14] Drying often becomes easier as the temperature approaches T_g' or the eutectic temperature: the sublimation rate can increase by about 13% for each 1°C rise in temperature.[20] Resistance to drying also decreases with decreasing product thickness (i.e. filling height) and with increasing vial diameter (which influences the area of the drying surface).[20] It is important that the temperature of the actual product remains constant (and therefore is monitored) throughout primary drying. The sublimation rate will decrease as primary drying proceeds and therefore the degree of product cooling due to sublimation will decrease also. The purpose of shelf warming is to counter this sublimation cooling. Accordingly, be sure to adjust the heat input to the shelves as the product dries to prevent a net rise in product temperature. An unchecked rise in product temperature could lead to collapse. One can use chamber pressure to control the length of the primary drying cycle.[20] Even under uniform conditions, primary drying times may vary by up to 10% for a given process. Make sure to include a delay period (ascertained empirically and occupying a period of hours) at the end of the primary drying cycle to ensure that all ice has sublimed satisfactorily and to avoid collapse.[20]

Secondary Drying

Secondary drying removes the remaining unfrozen, bound water from the lyophilizing material yielding a final product, low in residual moisture, which will be stable for an extended period without deterioration. The unfrozen water is removed by heating the shelves on which the primarily-dried product rests. The rule of thumb for initiation of secondary drying is the equivalence of product and shelf temperatures. For safety, however, one should include a delay period at the end of primary drying in any freeze drying protocol to prevent collapse of any vials which have not quite finished sublimation.[20] The partial pressure of water within the drying chamber drops at the end of primary drying as the last of the ice sublimes. If it can be monitored, this drop in water partial pressure can be a good indicator of the completion of primary drying. Even during secondary drying, with much of the original water gone, the preparation's temperature should never rise above T_g', something which is seldom appreciated.[14] T_g', however, rises as the residual water content drops.[20] One can, therefore, increase the temperature (within limits) during secondary drying. The vacuum need not be exhaustive during secondary drying; indeed, one should use a pressure in the region of 0.2 Torr for secondary drying.[20] There is little evidence that one can harm a protein by overdrying it.[20] However, some proteins, at least, may not possess the full native conformation after freeze-drying. Desai et al examined the effects of lyophilization on the structure of bovine pancreatic trypsin inhibitor (BPTI), a stable protein with a well-characterized folding pattern. These workers used hydrogen isotope exchange, detected by high resolution NMR, to monitor BPTI conformation in the lyophilized and rehydrated states. This technique has the advantage that any changes observed can be localized to particular amino acid residues. In BPTI, six hydrogen-bonded amide protons located within the protein interior (Tyr21, Phe22, Tyr23, Phe33, Tyr35, Phe45) show negligible rates of exchange with solvent in aqueous solution, even over a period of days or longer. Deuterium-labeled BPTI was lyophilized from a deuterated aqueous solution and then exposed to ordinary water vapor (67% relative humidity) under controlled conditions for 24 hours. After this time the solid BPTI was redissolved in the same deuterated buffer from which it had been freeze-dried and its proton NMR spectrum determined. Any deuterium-hydrogen exchanges noted in the BPTI molecules could have occurred only during exposure of the lyophilized sample to water vapor. Isotope exchange had occurred at all six of the amide positions indicated above, suggesting that these residues were not confined to the BPTI interior during freeze drying (or in the freeze-dried state), but rather had been exposed to the water vapor. Only a (partial) unfolding event could have brought this about. This was confirmed by a further experiment in which deuterated BPTI was lyophilized in the presence of the lyoprotectant sorbitol, which suppresses unfolding (see above and chapter 5). No isotope exchange took place under these conditions. These findings indicate that the freeze dried BPTI exists in a partially reversibly unfolded state.[27] They also underline the importance of using appropriate additives when performing lyophilization.

Freeze Drying Equipment

The most basic lyophilizer equipment will consist of a condenser/vacuum pump unit to which one may attach a centrifugal test tube holder or a manifold for the drying of multiple round-bottomed flasks. Freezing is accomplished in a separate cooling bath, usually filled with an alcohol. Such equipment can be used successfully for small-volume samples but fine and reproducible control of the overall process may not be possible. Neither can one attain continuous recording of sample temperature. Higher-grade equipment, with temperature-programmable shelves and a number of temperature probes, is greatly preferred. Shelf-equipped freeze dryers are especially suitable for use with rubber-capped pharmaceutical vials or with microtiter plates. Often an externally-operated screw press will be present, allowing one to seal vials (partially closed with rubber stoppers so as not to restrict gaseous movement) under vacuum before releasing air into the chamber. Obviously one should thoroughly familiarize oneself with the machine to be used. The user should know the locations of the main components and indicators together with appropriate settings for the meters, controls and valves (see Fig. 9.4).

The characteristics of the vials (usually glass) in which product is lyophilized will influence the process. The vial diameter, glass type and bottom shape and thickness will all affect the rate of heat transfer from the shelves to product. The vials should withstand freezing and pressure changes and should be uniform with respect to internal diameter and bottom thickness. The vial bottoms should be completely flat to make good contact with the shelves. The same model of vial should be used consistently as a change of vial type will introduce a variable into an optimized process.[20] Ensure that any other vessels used (e.g. round bottomed flasks) are of sufficient quality to withstand the temperatures and pressures associated with freeze drying. The directions for freeze drying given in the paragraphs below are of necessity very general. Freeze dryers differ greatly in their specifications depending on the model or manufacturer. Exact details will depend on the material and apparatus in question and on the user's requirements. Regardless of the type, one must use it at all times in accordance with the manufacturer's instructions. The method set out below assumes that the reader also studies the user's handbook. In pilot or manufacturing operations, Good Manufacturing Practice, ISO 9000 or regulatory authorities will likely impose many procedural disciplines which are beyond the scope of this chapter.

Start-Up

Ensure that the valve connecting the vacuum to the drying chamber is closed. Start the vacuum pump and allow it to evacuate. Observe the decrease in pressure on the vacuum indicator. It is important to start the pump first so as to reach a steady state high vacuum long before evacuation of the main chamber. It is also important that the pump warms up thoroughly before the condenser or shelves are cooled: water in any form reaching the pump can cause damage. A warm-up time of 30 minutes will usually suffice. Close the condenser drain outlet, which should always be left open when the freeze dryer is not in use. (If not, open it, allow any water to drain completely and then close tightly once again.) Switch on the condenser and allow to cool to -60°C. At this point, it may be convenient to cool the shelves slightly: this will aid the overall chilling process later. Do not cool the shelves more than a few degrees below ambient temperature, however, to prevent atmospheric condensation. Condensation will add to the water load which must be removed during the lyophilization process.

Filling and Loading

Now loading of the shelves with the preparation to be lyophilized can begin. (The volumes of bulk liquid subjected to freeze drying must never exceed the manufacturer's recommendations. If the condenser's ice capacity is reached or exceeded, the degree of product drying will be insufficient and many problems can result.) Fill only minimal amounts of material into each container to ensure a high ratio of surface area to volume: this will aid effective freeze drying. The product matrix will tend to inhibit sublimation of water vapor from the surface of the ice crystal. This resistance depends on the depth of liquid and on the solids content of the product.[20] Also, a good head space in the vial or ampoule will allow easier and better gaseous movement. One can conveniently load filled vials into metal trays at the bench and then place the trays into the drying chamber. The trays should have level bottoms to make good contact with the freeze dryer shelves. Many vials have narrow necks which one closes with special rubber stoppers. The design of these stoppers allows one to cap the vials partially while maintaining contact between their contents and the atmosphere. (Sealing of the vials takes place later under vacuum.) If one wishes to use such closures, one should partially stopper the vials before loading them into the freeze dryer. If using microtiter plates, place them within special metal frames which then come into contact with the shelves. The freeze dryer will likely have a number of flexible temperature probes; carefully clean each of these. Place some of the probes so as to monitor the temperatures of the shelves. Dip others directly into the material to be lyophilized. A temperature record of the actual solution (as distinct from the shelf underneath it) is well worth the loss of a small number of product vials. Arrange the temperature probes in different locations throughout the chamber: in the centers of shelves and also at the edges. Conditions will not be homogeneous across all vials and monitoring should be as complete as the equipment will allow. Insert the temperature probes right to the bottom of vials. Ensure that the probes remain in contact with the intended sample or shelf and do not become displaced during product loading or manipulation, or during lyophilization.

The Freeze Drying Operation

Once loading is complete, close and secure the door of the drying chamber. Now the freezing process may begin. Bring the material well below its freezing temperature (or glass transition temperature) as quickly as possible. Allow the shelf/product temperature to fall further to a steady -40°C before drawing the vacuum. Check that the drying chamber door is properly closed and sealed, then evacuate the chamber by opening the vacuum valve. The vacuum gauge reading will indicate atmospheric pressure as the valve is opened and may take a few minutes to show a vacuum once again as the air within the drying chamber evacuates. Soon, however, the pressure within the drying chamber will decrease and a steady state high vacuum will result. There now exists a high vacuum within the drying chamber together with a temperature gradient from the shelves/product (-40°C) to the condenser (-60°C). These conditions permit sublimation of the bulk water in the product over a period of hours as the shelves are gently heated. The sublimed water will collect as ice on the condenser. Sublimation, or primary drying, removes only the bulk water in the system; it is insufficient to remove the 'bound' water closely associated with the protein molecules. Removal of this bound water, or secondary drying, requires heating to elevated temperatures. This is usually applied through the shelves on which the product rests (and has previously been frozen). One can program the shelves to heat to a particular temperature at a defined rate appropriate for the product undergoing lyophilization.

Termination of Run and Removal of Product

One may terminate the freeze drying process when the cake has a good appearance and the product has reached a sufficiently low steady state percentage moisture (determined in a separate series of experiments). Seal vials, partially capped with rubber stoppers before insertion into the chamber, by operating the special screw press. This operation will insert the stoppers fully before air is admitted into the drying chamber. Sealing of vials while still under vacuum has an important advantage over the use of an inert, moisture-free gas such as nitrogen. Air will rush in when a vial, sealed under vacuum, is uncapped. The sound of the in-rushing air will be absent from a defective vial which has failed to seal or where the seal has broken down. In this way the user will immediately know of the defect and be aware that the vial contents may have deteriorated.[19] Close the valve to the vacuum pump firmly. Slowly release the vacuum within the drying chamber by a minimal opening of the air inlet valve: air admission should be as gentle as possible to prevent undue sudden stresses on sealed vials and also to prevent disturbance or upset of chamber contents by a strong jet of incoming air. The vacuum/pressure gauge will show a rise in drying chamber pressure; this will eventually equalize to atmospheric and one can then open the chamber door. Vials of freeze-dried product can be removed for storage (preferably at refrigerated temperatures). Store microtiter plates immediately in a desiccator or (better) vacuum-sealed within foil packets containing desiccant sachets.

Shutting Down

If the freeze dryer is to be re-used immediately, one must be certain that the condenser's ice capacity will withstand the accumulation of ice from two runs. If there is no more material for lyophilization, shut down the apparatus carefully. Remove any material spilled in the drying chamber and clean the shelves carefully according to the manufacturer's instructions. Leave the chamber door slightly ajar to allow circulation of air and to prevent sticking and compression of the door seals. Switch off the condenser and open its drain outlet. (The drain should remain open until the freeze dryer is next used.) Over a period of hours, the ice on the condenser surfaces will melt and drain away. Once the condenser temperature has returned to ambient and all the melted ice has drained away, allow the vacuum pump to run for a further 3 hours before shutting it down. This is to prevent any damage to the pump due to occurrence or accumulation of condensation. To maintain good pump performance, change the oil often according to the handbook's directions.

Using Simpler Freeze Dryers

Use of simpler apparatus with manifold or centrifuge accessories is carried out in much the same stepwise fashion as above. In these cases, aliquots of the product are frozen in individual open-necked flasks or tubes. Switch on the vacuum pump and condenser and allow to run as noted above. Freeze flask contents by immersion in an alcohol cooling bath or in liquid nitrogen (follow the normal precautions). Swirl or rotate the flask during the freezing step to effect even distribution of the product over the widest possible surface area. This will minimize the depth of material through which water loss must occur.[20] Connect the frozen material directly to the manifold assembly (or load into the centrifugal tube dryer) and draw a vacuum in the chamber as quickly as possible so as to minimize back-melting of ice under atmospheric pressure. The rapid reduction of pressure by the vacuum pump will aid sublimation and help minimize melting to water. While control of temperature and heating rates are more problematic with such accessories, visual inspection of the material and of the dried cake is much easier than in an enclosed chamber.

Quality Indices

Most freeze dryers will have a built-in chart recorder to provide a profile of shelf temperatures (and perhaps other parameters such as vacuum) during the lyophilization run. Study this carefully and retain with other batch information. Note the cake shape and texture and any variations between vials, especially those located at different positions within the chamber. Choose a representative number of vials from different chamber locations for scrupulous moisture determination. Shelf location is an important parameter in ensuring consistent sampling for determination of moisture.[28] The residual moisture content of, and its distribution throughout, the lyophilized preparation will dictate its long-term stability. Uneven moisture distribution between vials often leads to biphasic activity-loss profiles on extended storage.[14] Each 1% of moisture can depress T_g' by more than 10°C.[15] Jaffe et al investigated a gradual increase in residual moisture in vials containing small amounts of freeze dried thrombin. (Luckily, the increased moisture did not seem to affect thrombin potency.) Autoclaved, dried stoppers seemed to retain some moisture and lyophilization removed only surface, but not internal, moisture from the stoppers. Slow equilibration of moisture through the stoppers was thought to give rise to the gradual increase in residual moisture. The problem was solved by adopting a special drying treatment for vial stoppers and by changing the freeze drying cycle where vials contain small quantities of thrombin. Two holding times within the secondary drying portion of the run were lengthened so as to increase moisture removal from both the stopper and the thrombin cake.[28]

One should accurately measure the yield or recovery (%) of the initial biological activity by appropriate assay following rehydration of a representative number of vials. While waiting to assay, note the time required for complete product rehydration. Also note whether any turbidity remains on rehydration, or after what interval turbidity appears in a clear sample.[14] Persistence of the rehydrated biological activity can be measured at suitable or convenient time intervals. Accelerated degradation methods (chapter 1) can predict shelf lives of the lyophilized preparation at temperatures of interest for long-term storage.

Note that the formulation used greatly influences yield while the process parameters affect ease of rehydration and shelf life.[14] The glass transition temperature, T_g', and the product water content are critical parameters in the freeze drying process. It is essential to prevent or minimize damage to the protein of interest during freezing. Inclusion of excipients (additives) with high T_g' values in the protein formulation to be freeze-dried can be very useful. The mixture will form a glass at relatively low temperatures, minimizing freezing damage. A high T_g' will also allow the use of higher temperatures during primary drying with less danger of product collapse. Any constituent that will lower the unbound water content of the freeze concentrate will help shorten the secondary drying operation[14] but uncrystallized salts will decrease T_g' since any salt will bring about a depression of freezing point. Thus the salts content of the product formulation should be as low as is practicable.[14] The optimum temperature for freezing and primary drying depends on the ratio protein: protectant additive: salt in the freeze concentrate rather than in the initial solution.[14,21] The ratio protein: other solids in the freeze concentrate influences T_g' (ref. 14).

References

1. Mozhaev VV. Mechanism-based strategies for protein thermostabilization. Trends Biotechnol 1993; 11:88-95.
2. Sluzky V, Klibanov AM, Langer R. Mechanism of insulin aggregation and stabilization in agitated aqueous solutions. Biotech Bioeng 1992; 40:895-903.
3. Liu WR, Langer R, Klibanov AM. Moisture-induced aggregation of lyophilized proteins in the solid state. Biotech Bioeng 1991; 37:177-184.
4. Beynon RJ, Bond JS, eds. Proteolytic Enzymes: a Practical Approach. Oxford: IRL Press, 1987.
5. Scopes RK. Protein Purification: Principles and Practice. 3rd ed. Berlin: Springer Verlag, 1994:317-324.
6. Erlanger BF, Kokowsky N, Cohen W. The preparation and properties of two new chromogenic substrates of trypsin. Arch Biochem Biophys 1961; 95:271-278.
7. Schein CH. Solubility as a function of protein structure and solvent components. Biotechnology 1990; 8:308-317.
8. Reid TW, Wilson IW. *E. coli* alkaline phosphatase. In: Boyer PD, ed. The Enzymes. 3rd ed, vol. 4. London: Academic Press, 1971:373-415.
9. Carpenter JF, Crowe JH. The mechanism of cryoprotection of proteins by solutes. Cryobiology 1988; 25:244-255.
10. Hatley RHM, Franks F, Mathias SF. The stabilization of labile biomolecules by undercooling. Process Biochem 1987; 22:169-172.
11. Franks F. Storage stabilization of proteins. In: Franks F, ed. Protein Biotechnology: Isolation, Characterization and Stabilization. Totowa, NJ: Humana Press Inc, 1993:489-531.
12. More N, Daniel RM, Petach HH. The effect of low temperatures on enzyme activity. Biochem J 1995; 305:17-20.
13. Daniel RM, Dines M, Petach HH. The denaturation and degradation of stable enzymes at high temperatures. Biochem J 1996; 317:1-11.
14. Franks F. Freeze drying: from empiricism to predictability. Cryo-Letters 1990; 11:93-110.
15. Franks F, Hatley RHM, Mathias SF. Materials science and the production of shelf-stable biologicals. Pharm Technol Int 1991; 3:24-34.
16. Franks F, Hatley RHM. Storage of materials. US Patent no. 5,098,893: 1992.
17. Gibson TD, Woodward JR. Enzyme stabilization. World Patent no. PCT/GB91/00443, Publ. No. W091/14773: 1991.

18. Gibson TD, Higgins IJ, Woodward JR. Stabilization of analytical enzymes using a novel polymer-carbohydrate system and the production of a stabilized, single reagent for alcohol analysis. Analyst 1992; 117:1293-1297.
19. Thuma RS, Giegel JL, Posner AH. Manufacture of quality control materials. In: Howanitz PJ, Howanitz JH, eds. Laboratory Quality Assurance. NY:McGraw-Hill, 1987:101-123.
20. Pikal MJ. Freeze-drying of proteins. Part 1: process design. BioPharm 1990; 3:18-27.
21. Pikal, M.J. Freeze-drying of proteins. Part 2: formulation selection. BioPharm 1990; 3:26-30.
22. Perutz MF. Protein function below 220 K. Nature 1992; 358:548.
23. Franks F, Murase N. Nucleation and crystallization in aqueous systems during drying: theory and practice. Pure Appl Chem 1992; 64:1667-1672.
24. Hanson MA, Rouan SKR. Formulation of protein pharmaceuticals. In: Ahern TJ, Manning MC, eds. Stability of Protein Pharmaceuticals, Part B: In Vivo Pathways of Degradation and Strategies for Protein Stabilization. NY: Plenum Press 1992:209-233.
25. Blanchard JS. Buffers for enzymes. Meths Enzymol 1984; 104:404-415.
26. Timasheff SN. Stabilization of protein structure by solvent additives. In: Ahern TJ, Manning MC, eds. Stability of Protein Pharmaceuticals, Part B: In Vivo Pathways of Degradation and Strategies for Protein Stabilization. NY: Plenum Press 1992:265-285.
27. Desai UR, Osterhout JJ, Klibanov AM. Protein structure in the lyophilized state: a hydrogen isotope exchange/NMR study with bovine pancreatic trypsin inhibitor. J Am Chem Soc 1994; 116:9420-9422.
28. Jaffe DR, Samyn JC, Sheehan G. Techniques used to freeze-dry "small doses" of thrombin. Develop Biol Stand 1991; 74:211-214.

CHAPTER 10

Conclusion

These chapters have tried to show that many different approaches have contributed to our understanding of, and ability to manipulate, protein stability. It is likely that such cross-fertilization will continue in the future. Advances in determining three-dimensional protein structures will aid molecular modeling and rational design approaches. Mutagenesis and expression techniques continue to improve while chemical techniques are likewise becoming more sophisticated. It is likely that these powerful and complementary strategies will be increasingly combined in the future. Indications of this have already emerged in work such as the combined mutational/chemical stabilization of cytochrome c[1] and of subtilisin.[2]

This book has surveyed the features of proteins that dictate their stability characteristics and has reviewed a number of alternative strategies for increasing the in vitro stability of a target protein. The study of thermophilic proteins has yielded much information on those features and interactions which stabilize protein structure. It has also provided many examples of ready-made stable enzymes which may be directly applied for a number of purposes. Denaturation studies on native enzymes have indicated the molecular changes leading to loss of function. One must prevent or limit these types of changes in order to stabilize a target protein. Enzyme catalysis in organic solvents has revealed some unexpected features which have important stability implications. These three strands of enzymology help greatly in the formulation of coherent strategies of enzyme stabilization.

Available stabilization strategies include immobilization, use of additives, chemical modification and protein engineering. Each of these has been discussed at length. Protein engineering is by far the most novel of these: the others have been practiced for many years. The development of protein engineering had to await the arrival of convenient modern techniques of oligonucleotide synthesis and site-specific mutagenesis, since it offers its greatest potential where the target protein's full amino acid sequence and exact three-dimensional structure are known. Methods such as Laue X-ray techniques and phosphorothioate nucleotide mutagenesis (which eliminates the need to select for recombinants) have been useful additions to the protein engineer's armory. Chemical modification can act as a useful complement to protein engineering, as pointed out in chapter 8 and in refs. 3 and 4. It can also yield stabilized proteins in its own right, by molecular crosslinking or by modification of specific functional groups. Immobilization and solute/ additive use have been well characterized in terms of their stabilizing effects but they may not be suitable for certain applications.

Stabilizing Protein Function, by Ciarán Ó Fágáin.
© Springer - Verlag and Landes Bioscience 1997.

Table 10.1. Comparison of protein stabilization strategies

Technique	Requirements	Applications/ Suitable for...	Merits	Drawbacks
Immobilization	Activated solid phase, multiple protein-solid phase links	Bioreactors (column or batch), biosensors, diagnostics	Widely practiced, abundant literature, many supports available (often activated by various chemistries)	Possibility of: activity losses, poor orientation, steric hindrance, altered kinetics
Additives	A range of osmolytes, preservatives, etc.	Low-/ moderate- temperature and/ or dried/ freeze-dried storage	Avoids actual alteration to target protein, significant stabilization possible, possible formulation of thera-peutics using only approved excipients	Possible loss of stabilizing effects on reconstitution, dilution or buffer exchange

Chemical modification	Suitable active reagents, including (but not confined to) crosslinkers	Many in vitro applications, complements protein engineering,[a] applicable to recombinant proteins[b]	Direct alteration of target protein molecule	Regulatory approval perhaps difficult for therapeutics
Protein engineering	3-D structure (usually), cloned gene, suitable molecular biological/ mutation techniques, appropriate selection system	Many applications in vitro and in vivo	Permanent alteration of protein stability at molecular level	Regulatory requirements may be rigorous

Note that it is possible to combine some of these techniques, e.g. alanine aminotransferase (ALT) modified with dimethylsuberimidate and immobilized on adipic acid dihydrazide-agarose was more stable than ALT which had been either modified or immobilized alone.[c] Muheim and colleagues combined protein engineering and chemical modification to stabilize cytochrome c: a metal-mediated crosslink resulted from reaction of a ruthenium-bipyridyl compound with the natural (wild type) His39 and the engineered His58 substitution.[d]

a. Profy AT, Schimmel P. Complementary use of chemical modification and site-directed mutagenesis to probe structure-activity relationships in enzymes. Prog Nucl Acids Res & Mol Biol 1988; 35:1-26.

b. Smith RAG, Dewdney JM, Fears R et al. Chemical derivatization of therapeutic proteins. Trends Biotechnol 1993; 11:397-403.

c. Moreno JM, Ó Fágáin C. Stabilization of alanine aminotransferase by consecutive modification and immobilization. Biotechnol Lett 1996; 18:51-56.

d. Muheim A, Todd RJ, Casimiro DR et al. Ruthenium-mediated cross-linking and stabilization. J Am Chem Soc 1993; 115:5312-5313.

It is likely that each of these stabilizing strategies will find continued use. It is also likely that combinations of the available techniques will become increasingly important. For example, one can envisage a thermophilic enzyme being chemically altered to enhance its stability before being immobilized to increase its stability still further. Already, a dimethylsuberimidate-modified, stabilized variant of alanine aminotransferase (a clinical marker of liver function and a potential catalyst in amino acid synthesis) has been stabilized further by subsequent immobilization.[5] Protein engineering has been used for thermostabilization but, as pointed out by Arnold,[6] one may use it also to tailor an enzyme for use in hydrophobic solvents, which use may be thermostabilizing in itself. Arnold's group has reported the 'evolution' of a useful 'organotolerant' esterase through successive generations of random mutagenesis and selective screening. Interest in this case focused on activity rather than on stability.[7] The antibiotic loracarbef, a fourth-generation cephalosporin, is synthesized in protected form as its *p*-nitrobenzyl ester which then undergoes deprotection in aqueous-organic medium to yield the final product. The esterase enzyme (a *B. subtilis* enzyme expressed in *E. coli*) can conveniently be used for this (non-natural) deprotection reaction. One mutant displayed a 16-fold increase in esterase activity over that of the wild type enzyme and was as active in 30% dimethylformamide as was the wild type in aqueous medium. Two positive variants were subjected to random pair-wise gene recombination and this in turn yielded a further 2-fold improvement in activity.[7] The combination of random mutagenesis with appropriate selection (molecular Darwinism?) will likely prove a fertile strategy for the development of more stable enzymes and proteins also (see chapter 8 for a review of one case history to date, subtilisin stabilization by random mutagenesis).

Other 'unusual' stabilizing options may include the use of high hydrostatic pressures. Mozhaev et al have described how both activity and stability of alpha-chymotrypsin were increased by high pressures up to a critical value, beyond which pressure-induced denaturation occurred. The activation effect increased with both pressure and temperature: at 50°C and 3.6 kbar, for example, the activity of chymotrypsin was 30 fold higher than at 20°C and 1 bar. Pressures up to 1.2 kbar increased chymotrypsin's heat stability; however, pressures greater than 3.6 kbar (at 50°C) denatured the enzyme. At 55°C and 1.8 kbar, chymotrypsin remained active for quite some time, while losing activity almost immediately at the same temperature and only one atmosphere pressure. Inclusion of 40% glycerol led to a further gain in stability at 1.8 kbar: T_m increased by a further 5°C. 40% Glycerol also exerted a protecting effect against higher, denaturing, hydrostatic pressures. The activation effect arises from changes in the activation free volume ($\Delta V^{\neq}$), a quantity analogous to the activation Gibbs energy, $\Delta G^{\neq}$. Chemical reactions proceeding with a negative change in activation volume will be accelerated by higher pressures.[8] At first sight it seems remarkable that the same phenomenon—high pressure—can act as both a stabilizing and a denaturing influence but the differing effects are a function of pressure. Moderate pressures can dissociate oligomeric proteins while somewhat higher pressures denature small monomeric proteins, so there is an upper limit to the pressures that enzymes can endure.[8] A detailed discussion of these pressure-induced enzyme activation/stabilization phenomena may be found in refs. 8 and 9. Many industrial processes take place under pressures greater than atmospheric and the possibility of running enzyme-catalyzed processes for reasonable time periods at moderate pressures will likely interest industrial chemists. It is likely that high-pressure enzyme studies ('hyperbaric enzymology') will become more plentiful in the future.

A tremendous growth has taken place in our understanding of enzyme stability and its possible enhancement. One can now plan a stabilization strategy on a rational basis using one of (or a combination of) the options outlined in this volume. Of course, there is no guarantee that a particular procedure, no matter how well chosen a priori, will yield a stabilized enzyme derivative. However, the number of unsuccessful attempts should be greatly reduced, since a process of elimination will suggest further factors to take into account towards a successful outcome.

Towards an Exciting Future?

In the future, it is likely that stabilized enzymes will become available for high-value products and processes, where the development costs will be more easily justified and recouped than in the case of low-cost or price-sensitive 'commodity' products. (Even in these cases, however, a stabilized biocatalyst could give rise to immediate cost savings.) There may soon come a new wave of biotechnological advance spearheaded by the application of purposely-stabilized enzymes. Combination of different protein domains could give rise to an extra stabilizing framework or cage surrounding the active site portion of the protein. Active sites of enzymes may be altered in terms of geometry or constituent residues to optimize activity for a given application. Active site domains could also be combined to yield a single protein with two distinct catalytic activities. Already Mosbach's group has created bifunctional enzymes by gene fusion, allowing the catalysis of sequential reactions in an industrial process, metabolic pathway or coupled enzyme assay in an analytical device.[10] This represents a more sophisticated version of older technologies (co-immobilization and chemical crosslinking[11]) for bringing two or more enzymes into close proximity. Structural genes for the two enzymes of interest can be ligated together and the resulting chimeric gene will encode both active sites on a single polypeptide. (Some care and reasoning must be exercised in each case to ensure a feasible fusion strategy, of course.) Fusions need not be confined to monomeric enzymes. Examples of bifunctional enzymes prepared to date include beta-galactosidase-galactokinase, beta-galactosidase-galactose dehydrogenase, galactose dehydrogenase-luciferase and even a trifunctional beta-galactosidase-galactose dehydrogenase-galactokinase construct. (Beta-galactosidase is a tetramer and galactose dehydrogenase is a dimer.) The kinetic properties of each enzyme in the fusion product can be influenced by alterations to the reaction medium. The luciferase enzyme is of special interest. This heterodimer catalyzes a luminescent reaction useful for quantifying NAD^+ levels (see also chapter 8). It is very prone to inactivation by immobilization or chemical modification procedures, so gene fusion represents the most feasible means of coupling it to other enzymes of interest. In general, short linkers of 2-10 residues length have been found optimal for fusion; longer linkers are prone to proteolysis, reducing yields.[10] In similar manner, Terashima et al have described the construction of a chimeric enzyme from two isoenzymes of rice alpha-amylase. While one of the constituent isoenzymes was more active in soluble starch hydrolysis and the other was better at degrading oligosaccharides, the chimera was highly active in both applications. Thermostability characteristics of the chimera resembled those of the oligosaccharide-active constituent isoenzyme.[12] In this way, one could develop a purpose-built in vitro multienzyme complex. One could also design stabilized or altered antibodies for conjugate, imaging or therapeutic use. Stabilized enzymes, antibodies and receptor proteins would all be applicable to biosensor development, perhaps allowing a new generation of versatile, reusable biosensors to act under the sorts of rigorous conditions which have so far prevented their widespread use.

These developments will be aided by parallel advances in protein purification and folding and in recombinant expression techniques. There have been many examples to date of fusion proteins where a protein of interest is recombined with an easily-detectable or easily-purified carrier protein. Fusions can be designed for convenient affinity purification from an impure mixture. This step is followed by specific cleavage of the bond linking the carrier with the target protein, facilitating isolation of the latter. More recently, fusions have been created to assist the target protein's folding in vivo. Samuelsson and colleagues reported increased yields of human insulin-like growth factor (IGF-I) following coexpression of a specific binding protein (IGFBP-1) in *E. coli.* Use of a glutathione redox buffer during growth permitted formation and breakage of disulfide bonds in the host cell periplasm. Both the hormone and the binding protein were produced as affinity fusions and this approach allowed the choice of alternative affinity methods for purification of the heterodimer formed in vivo.[13]

Correct glycosylation of recombinant proteins is often essential for therapeutic use in vivo: incorrect or incomplete glycosylation of recombinant erythropoietin (EPO) results in its rapid degradation without any therapeutic effect even though its polypeptide portion is correctly folded.[14] Often mammalian cell lines such as baby hamster kidney (BHK) or Chinese hamster ovary (CHO) cells must be used for expression, with all the attendant cost and culture considerations. Baculovirus expression vector (BEV) insect cell systems combine high yields (> 30 $mg.l^{-1}$) with short process development times (a few weeks) and can perform many post-translational processing steps. However, these systems' glycosylation abilities do not compare favorably with those of mammalian cell lines. The most commonly used insect cell line is *Spodoptera frugiperda* but Ogonah et al decided to explore the use (and glycosylation capabilities) of cells from a different species, *Estigmena acrea*, in a BEV system. Recombinant human gamma-interferon was the expression protein studied; this has two potential N-linked glycosylation sites. The *E. acrea* cells performed much more extensive glycosylation of the interferon, with greater production of hybrid or complex type N-glycans, than the *S. frugiperda* cells.[15] Further work in this area may throw up more high-yielding and cost-effective expression systems capable of 'correct' glycosylation of recombinant proteins.

Even more fascinating things may lie ahead. Drexler has reviewed the emerging topic of nanotechnology and discusses ways in which proteins or protein-like molecules may be 'improved' for use as nanomachines.[16] Peptide-based, self-assembling nanoscale tubular structures have been reported with possible applications in chemistry, catalysis, molecular electronics and separation technology.[17] In addition to a number of the stabilization strategies suggested in this volume, Drexler suggests the synthesis of *branched* polypeptide molecules,[16] based on cited work of Mutter and others.[16,18] Such a strategy would always demand quite some chemical ingenuity, as no branched peptide or polypeptide could result directly from the use of recombinant techniques (at least, as currently understood). However, advances in peptide chemistry and in enzymatic peptide synthesis[19] (including 'chemoenzymic' synthesis[20]) are continuing and these could be instrumental in solving these sorts of problems.

References

1. Muheim A, Todd RJ, Casimiro DR et al. Ruthenium-mediated cross-linking and stabilization. J Amer Chem Soc 1993; 115:5312-5313.

2. Grøn H, Bech LM, Branner S et al. A highly active and oxidation-resistant subtilisin-like enzyme produced by a combination of site-directed mutagenesis and chemical modification. Eur J Biochem 1990; 194:897-901.
3. Estell DA. Engineering enzymes for improved performance in industrial applications. J Biotechnol 1993; 28:25-30.
4. Profy AT, Schimmel P. Complementary use of chemical modification and site-directed mutagenesis to probe structure-activity relationships in enzymes. Prog Nucl Acid Res & Mol Biol 1988; 35:1-26.
5. Moreno J, Ó Fágáin C. Stabilization of alanine aminotransferase by consecutive modification and immobilization. Biotechnol Lett 1996; 18:51-56.
6. Arnold FH. Enginering enzymes for non-aqueous solvents. Trends Biotechnol 1990; 8:244-249.
7. Moore JC, Arnold FH. Directed evolution of a *para*-nitrobenzyl esterase for aqueous-organic solvents. Nature Biotechnol 1996; 14:458-467.
8. Mozhaev VV, Lange R, Kudryashova EV et al. Application of high hydrostatic pressure for increasing activity and stability of enzymes. Biotechnol Bioeng 1996; 52:320-331.
9. Mozhaev VV, Heremans K, Frank J et al. High pressure effects on protein structure and function. Proteins: Struct Func Genet 1996; 24:81-91.
10. Bülow L & Mosbach K. Multienzyme systems obtained by gene fusion. Trends Biotechnol 1991; 9:226-231.
11. Rajput YS & Gupta MN. A conjugate of trypsin and chymotrypsin. Appl Biochem Biotechnol 1987; 16:201-210.
12. Terashima M, Kawai M, Kumagai MH et al. Characteristics of a chimeric enzyme engineered for two rice alpha-amylase isozymes. Appl Microbiol Biotechnol 1996; 45:607-611.
13. Samuelsson E, Jonasson P, Viklund F et al. Affinity-assisted in vivo folding of a secreted human peptide hormone in *Escherichia coli*. Nature Biotechnol 1996; 14:751-755.
14. Higuchi M, Oh-eda M, Kuboniwa H et al. Role of sugar chains in the expression of the biological activity of human erythropoietin. J Biol Chem 1992; 267: 7703-7709.
15. Ogonah OW, Freedman RB, Jenkins N et al. Isolation and characterization of an insect cell line able to perform complex N-linked glycosylation on recombinant proteins. Biotechnology 1996; 14: 197-202.
16. Drexler KE. Molecular nanomachines: physical principles and implementation strategies. Annu Rev Biophys Biomol Struct 1994; 23:377-405.
17. Ghadiri MR, Granja JR, Milligan RA et al. Self-assembling organic nanotubes based on a cyclic peptide architecture. Nature 1993; 366:324-327.
18. Mutter M & Vuilleumier S. A chemical approach to protein design—template-assembled synthetic proteins (TASP). Angew Chem Int Ed Engl 1989; 28: 535-676.
19. Wallace CJA. Peptide ligation and semisynthesis. Curr Opin Biotechnol 1995; 6:403-410.
20. Gaertner HF, Offord RE, Cotton R et al. Chemo-enzymic backbone engineering of proteins. J Biol Chem 1994; 269: 7224-7230.

Index

B

C

I

T

U

V

W

X

Y

Z